MW01087723

Purshia

JAMES A. YOUNG & CHARLIE D. CLEMENTS

Purshia: The Wild and Bitter Roses

University of Nevada Press
Reno & Las Vegas

University of Nevada Press, Reno, Nevada 89557 USA

Manufactured in the United States of America

The paper used in this book meets the requirements of American National Standard for Information Sciences—Permanence of Paper for Printed Library Materials, ANSI Z39.48-1984. Binding materials were selected for strength and durability.

FIRST PRINTING
11 10 09 08 07 06 05 04 03 02
5 4 3 2 1

Library of Congress Cataloging-in-Publication Data
Young, James A. (James Albert), 1937–
Purshia : the wild and bitter roses /
James A. Young and Charlie D. Clements.
p. cm.
ISBN 0-87417-491-0 (hardcover : alk. paper)
1. Purshia—Great Basin. I. Clements, Darin Duane, 1965– II. Title.
QK495.R78 .Y68 2002
583'.734'0979—dc21 2002000640

This work is dedicated to the two pioneers in bitterbrush research, August L. Hormay and Eamor C. Nord. Their research paved the way for future generations of scientists interested in the biology and management of native shrubs on rangelands.

Contents

List of Illustrations ix

List of Tables xi

Preface xiii

Chapter One

The Wild and Bitter Roses 1

Chapter Two

Hunters, Herdsmen, and Brush 14

Chapter Three

Bitterbrush Plant Communities 31

Chapter Four

Ecophysiology of *Purshia* 52

Chapter Five

Purshia Seed Physiology 73

Chapter Six

Seeding *Purshia* Species 98

Chapter Seven

Granivore Relations 124

Chapter Eight

Ruminant Nutrition 138

Chapter Nine

Insects and Plant Diseases 156

Chapter Ten

Wildfire Relations 171

Chapter Eleven

The Role of Nitrogen 195

Chapter Twelve

Purshia Management 211

Notes 229

Index 257

Illustrations

Figures

2.1. Bitterbrush 18
2.2. Apache plume 19
2.3. Antelope bitterbrush 21
2.4. Cliffrose 22
3.1. Woodlands with antelope bitterbrush understory 36
3.2. Landscape view from Janesville Grade 37
3.3. Jeffrey pine/desert bitterbrush community 42
3.4. Pine woodland with big sagebrush–antelope bitterbrush understory 44
3.5. Cattle grazing in western juniper/antelope bitterbrush–mountain big sagebrush/bluebunch wheatgrass community 46
4.1. Antelope bitterbrush in full flower 58
4.2. Height of antelope bitterbrush by age classes 70
4.3. Sizes and ages of five antelope bitterbrush plants 71
5.1. Cross section of antelope bitterbrush achene 74
5.2. Categories of seedbed temperatures used for comparison with germination profiles 96
6.1. Screen tray and paddle for harvesting antelope bitterbrush seed 99
6.2. Vacuum browse-seed harvester 100
6.3. Hoop and cloth bag for bitterbrush seed collection 101
6.4 Barley debearder for removing remnant style parts from achenes 103
6.5. Air screen for cleaning antelope bitterbrush seed 104
6.6. Normal and black shriveled antelope bitterbrush seeds 105
6.7. Root development of antelope bitterbrush seedlings 115
6.8. Antelope bitterbrush seeded in two strips 116
6.9. Hansen scalper–browse seeder 120
6.10. Schussler antelope bitterbrush spot seeder 121

7.1. Typical harvester ant mound 125
7.2. Rodents of the families Sciuridae, Heteromyidae, and Cricetidae 127
7.3. Fur-lined external cheek pouches typical of the family Heteromyidae 128
7.4. Antelope bitterbrush seedlings emerging from a scatter-hoard cache 128
7.5. Deer mouse (*Peromyscus maniculatus*) 131
8.1. Mixed community of mountain big sagebrush and antelope bitterbrush 145
9.1. Say's stinkbug (*Chlorochroa sayi*) 162
9.2. Great Basin tent caterpillar (*Malacosoma fragile*) 164
10.1. Bluebunch wheatgrass (*Pseudoroegneria spicata*) unburned in a wildfire fueled by cheatgrass 173
10.2.a. Juniper Hill ca. 1900 177
10.2.b. The same site in 1978 178
10.3. Periodicity of establishment of western juniper trees in low sagebrush communities 179
10.4.a. Frequency of fire scars on western juniper trees on low sagebrush sites 180
10.4.b. Cross section of a fire-scarred western juniper trunk 180
12.1. Heavily grazed rangeland, ca. 1900 212
12.2. Mule deer exclosure 216
12.3. A three-way big game exclosure built in the 1950s 217
12.4. Nearly senescent stand of antelope bitterbrush 219
12.5. Mule deer depredation of agricultural fields in northwestern Nevada 223
12.6. A lone antelope bitterbrush plant in a stand of big sagebrush with juniper 226

Maps

1.1. General distribution of antelope bitterbrush 5
1.2. Distribution of antelope and desert bitterbrush and cliffrose in western North America 6
1.3. Distribution of cliffrose and Apache plume in western North America 8

Tables

1.1. Common Names for *Purshia* Species 3
3.1. Foliage Cover for a *Juniperus/Artemisia-Purshia* Association 34
4.1. Seasonal Growth and Development of Bitterbrush 53
4.2. Phenology of *Purshia* Plants from Southeastern to Northeastern California 53
4.3. Cliffrose, Apache Plume, and Bitterbrush Accessions 54
4.4. Phenology of Cliffrose, Apache Plume, and Bitterbrush Accessions 56–57
4.5. Duration of Flowering and Fruiting in Bitterbrush, Cliffrose, and Apache Plume Accessions 60
4.6. Phenology of Antelope Bitterbrush in Oregon 61
4.7. Antelope Bitterbrush Seed Production 62
4.8. Sprouting of Antelope and Desert Bitterbrush Following Wildfires 65
4.9. Shrub Accessions Evaluated in the Baker, Oregon, Common Garden 67
4.10. Height of Shrub Accessions in the Baker, Oregon, Common Garden 68
5.1. Germination of Bitterbrush and Cliffrose Seeds Without Pretreatment 76
5.2. Germination of Antelope Bitterbrush Seeds after Cool-Moist Pretreatment 79
5.3. Germination of Antelope Bitterbrush Seeds after Cool-Moist Pretreatment in Sand 79
5.4. Mean Germination Profiles for Bitterbrush and Cliffrose Seeds 83
5.5. Germination Profiles for Bitterbrush and Cliffrose Seeds Incubated Without Pretreatment 85
5.6. Germination Profiles for Bitterbrush and Cliffrose Seeds Pretreated in Thiourea 85
5.7. Germination of Bitterbrush and Cliffrose Seeds Soaked in Thiourea 86–87

5.8. Germination Profiles for Bitterbrush and Cliffrose Seeds Pretreated with Moist Prechilling 88
5.9. Germination of Moist-prechilled Bitterbrush and Cliffrose Seeds 89–90
5.10. Germination Profiles for Bitterbrush and Cliffrose Seeds Soaked in Hydrogen Peroxide 91
5.11. Germination of Bitterbrush and Cliffrose Seeds after Soaking in Hydrogen Peroxide 92–93
5.12. Optimum Temperature Regimes for Germination of Bitterbrush and Cliffrose Seeds 94–95
5.13. Mean Germination for Four Seedbed Temperature Regimes and Treatments 97
6.1. Mortality of Bitterbrush Seedlings Emerging in the Spring of 1953 114
8.1. Digestible Nutrients of Browse Plants 143
8.2. Mule Deer Stomach Contents 146
8.3. Winter Crude Protein and Leafiness of Rose Family Shrubs 153
9.1. Insects Collected from *Purshia tridentata* 158–161
11.1. Soil Collection Sites in California 202
11.2. Nitrogen Characteristics of Soils Used in *Purshia* Nodulation and Nitrogen Fixation Experiments 203
11.3. Ethylene Production by Bitterbrush Seedlings 204
11.4. Nitrogen Fixation by Bitterbrush Grown in Soils with Different Fertility Treatments 205
11.5. Nodule Numbers of Cliffrose Grown in Soils with and without Nitrogen Enrichment 207
11.6. Antelope Bitterbrush Density and Height after One Growing Season 209
12.1. Mean Twig Growth of Bitterbrush Plants 224

Preface

More has been written about antelope bitterbrush than about any other shrub native to western North America. For more than a half-century antelope bitterbrush has been synonymous with deer management on ranges where big game animals seek food in the winter. The recent decline in the productivity of antelope bitterbrush stands and lack of recruitment of seedlings have prompted great concern in many states. The utilization of antelope bitterbrush browse by domestic livestock has become a major source of contention between ranchers and wildlife managers.

Large-scale projects have been developed to restore antelope bitterbrush stands. Such restorations have proven both expensive and difficult, however, and results have been inconsistent. This failure has driven research efforts to increase the understanding of *Purshia* seed and seedbed ecology. The tangled web of ecological interactions that has come into focus as the biology of this native shrub has been characterized reveals antelope bitterbrush to be symbolic of the twentieth-century western range. Nowhere is this more apparent than in the relation of antelope bitterbrush populations to wildfires, both in presettlement and historic times.

This book was written with multiple audiences in mind. In the broad view, it provides background for anyone interested in environmental issues. It reaches back to examine the vast environmental changes that occurred at the close of the Pleistocene and is as current as the vital range management issues of today. The detail of the biology presented meets the standards of scientists, yet the text is accessible to the lay reader. The comprehensive nature of the coverage provides natural resource managers, for the first time, with a basic reference for the species of *Purshia*.

We greatly appreciate the contributions of many colleagues and collaborators from the scientific, land and wildlife management, and ranching communities during nearly four decades of research on *Purshia* species. The late Eamor Nord influenced the direction of our research by encouraging us to continue and expand our work on antelope bitterbrush ecology and management. A special thank you to the host of undergraduate and graduate students who assisted with our project over the years.

Chapter One

The Wild and Bitter Roses

The road down Leadville Canyon from the volcanic highlands of northwestern Nevada to the Black Rock Desert is still surfaced with gravel, although it is greatly improved compared with 1910, when the underground mines were operating at Leadville. You get very familiar with the creek coming down the canyon because the road is continually forced from one side of it to the other by the burnt cliffs of volcanic debris. When you drive down the canyon in May, the numerous fords are no problem. Only the upper ones show any water, and it may be only a trickle flowing in the dark gravel. The uplands of northwestern Nevada are vast volcanic tablelands colored a uniform silver gray by sagebrush. The riparian vegetation in the bottom of the canyon is a welcome collage of gold and red willow stems and green leaves of cottonwoods that shimmer in the barest breeze. If you pause on a May day and walk along the canyon bottom, you will be rewarded with the delicate five-petaled flowers of wild roses (*Rosa woodsii* var. *ultramontana*). A rose in the desert?

The temperate deserts that dominate the landscapes between the Sierra-Cascade and Rocky Mountains are products of the post-Pleistocene aridity that grips the region. The western walls of the Sierra-Cascades cast a rain shadow across much of the Intermountain area. The rich mixed conifer and broadleaf forests and woodlands that once extended inland from the Pacific Coast have largely disappeared except in the higher mountain ranges.[1] In the Great Basin, the mixed conifer forests have disappeared even from most of the higher ranges, leaving only subalpine forests of five-needle pines. The lower elevations of the Great Basin feature pygmy conifer woodlands of pinyon (*Pinus* sp.) and juniper (*Juniperus* sp.) that invaded during the Holocene. The lowlands of the Intermountain area became dominated by shrub-steppe vegetation in a largely treeless environment at the same time.

Basin bottoms—the plains of Ice Age pluvial lakes—that have halomorphic (salt-affected) soils are sparsely clothed with shrubs of the chenopod family. Saltbushes (*Atriplex* sp.) dominate, but a host of sometimes diverse and sometimes monospecific genera has evolved here. No woody rose species could persist in the salt deserts, although legumes (*Psorothamnus* = *Dalea*) and even the rare woody Brassicaceae, bush peppergrass (*Lepidium fremonti*), are found in the arid basin bottoms. As the temperate deserts of the Intermountain area merge into the warm deserts of the Southwest, an entire life zone is characterized by the intricately branched woody rose blackbrush (*Coleogyne ramosissima*).

Vast areas of the foothills and mountain slopes are dominated by shrubs of the sunflower family, especially sagebrush (*Artemisia* sp.). The woody members of the rose family were never dominant in the Intermountain vegetation, but the fossil record indicates that they were significant understory woody plants in the forests and woodlands, much as they are now on the Pacific slope of the mountains. From these ancestor populations evolved a few species that play important roles in the current temperate desert environment because of their colorful spring flowers and for the nutritious browse they produce.

Some of the rose species persist in disjunct desert refuges. The interior wood rose in riparian habitats is an example. Oceanspray (*Holodiscus discolor*) is another relic woody rose that can be found in isolated patches on the north side of rocky spines on the higher ridgelines in the arid mountain ranges of the Great Basin. Oceanspray is also found as a shrub layer characterizing mixed conifer woodlands of the interior Pacific Northwest. Apparently this woody rose has inhabited these refugia since the Pleistocene.

Other woody rose species evolved in the post-Pleistocene aridity to form significant components of the western mountain and Intermountain vegetation; antelope bitterbrush (*Purshia tridentata*), desert bitterbrush (*P. tridentata* var. *glandulosa*), and cliffrose (*P. mexicana* var. *stansburiana*) are examples. Because they are still evolving through hybridization to meet the demands of changing environments, the classification of these three shrubs has changed, and probably will continue to be changed as we learn more about their origins. Desert bitterbrush has been given subspecies rank, and cliffrose was long known under the generic name *Cowania*.[2] These three plants usually occur as distinct, identifiable individuals, although sometimes the distinctions blur when separate populations come together and exchange genetic material. There is tremendous morphological variation within each of these woody rose family members, and great variation as well in the types of sites they inhabit. Despite this variation, each species or variety maintains a stereotypic image and each inhabits specific types of environments with sufficient frequency to characterize wildland sites. From British Columbia to Truckee, California, along the eastern slopes of the Cascade and northern Sierra Nevada, ponderosa (*Pinus ponderosa*) or Jeffrey pine (*P. jeffreyi*) woodlands with antelope bitterbrush understories have a distinct, repetitive, identifiable appearance that produces a characteristic landscape.

Antelope bitterbrush was first collected by Captain Meriwether Lewis on the "plains of the Columbia" during the Lewis and Clark Expedition to the mouth of the Columbia River.[3] The specimen made the long and perilous trip back up

TABLE 1.1.
Common Names for Purshia tridentata, P. tridentata *var.* glandulosa, *and* P. mexicana *var.* stansburiana

Purshia tridentata	*P. tridentata* var. *glandulosa*	*P. mexicana* var. *stansburiana*
antelope bitterbrush	desert bitterbrush	cliffrose
deerbrush	quinine bush	
buckbrush		
quininebush		
black sage		
greasewood		
antelope brush		
kunzia		

the Columbia River, over the Rocky Mountains, down the Missouri River, and all the way to Philadelphia, where it came into the hands of Frederick Traugott Pursh (1774–1820), a German-born botanist and naturalist-physician. After examining the botanical specimens returned by Lewis and Clark, Pursh published the *Flora Americae Septentrionalis* in 1814, describing antelope bitterbrush under the name *Tigarea tridentata.* Antelope bitterbrush remained hidden west of the mountains for much of the nineteenth century. Kurt Polykarp Joachin Sprengel (1766–1833) somehow received a specimen, however, and described it in Linnaeus's *Systema Vegetabilius* (vol. 2, p. 475 [1875]), which he edited from 1825 to 1828, under the name *Kunzea tridentata.* The species name, *tridentata,* describes the plant's three-lobed leaf tips. Augustin Pyramus de Candolle, a professor of botany at Geneva and one of the founders of phylogenetic classification, recognized that Pursh's description had priority and established the generic name *Purshia.* Thus, the current scientific name for antelope bitterbrush is *Purshia tridentata* (Pursh) DC. The genus name *Kunzea* nevertheless persisted well into the twentieth century and was sometimes even published as a common name for this shrub.

The number of common names assigned to a native plant species is a measure of its distribution and importance, and antelope bitterbrush certainly has accumulated its share (Table 1.1). The twigs and leaves of this shrub have a very bitter taste to humans, hence the widely repeated common name "bitterbrush." The bitterness is imparted by hydrocyanic acid, a compound found in many species of the rose family. *Purshia tridentata* is truly a wild and bitter rose.

Desert bitterbrush, *Purshia tridentata* var. *glandulosa,* was first collected in

1885 on the Mohave Desert side of Tehachapi Pass in southern California by Mary Katharine Curran (1844–1920), a physician-botanist who collected and described plants in the desert areas of the Southwest.

The genus *Cowania,* in which cliffrose was long classified, was named for James Cowan, a British merchant and amateur botanist who collected in South America and died in Lima in 1814. The specific name, *mexicana,* denotes the plant's general area of distribution in northwestern Mexico and the southwestern United States. The variety *stansburiana* honors the American topographical engineer–plant explorer Captain Howard Stansbury, who collected the type specimen on what is now Stansbury Island in the Great Salt Lake of Utah in 1849.[4] John Torrey (1796–1873), a physician and professor of chemistry and botany at the College of Physicians and Surgeons in New York City who described many of the plants collected by early explorers of the Far West, described the Stansbury specimens as *Cowania stansburiana.* Willis Linn Jepson, the noted California botanist, published the change to *Cowania mexicana* Don var. *stansburiana* Jepson in 1925. The specific name, *mexicana,* was originally published by David Don (1799–1841), a professor at King's College, London, and a librarian for the Linnaean Society. The description was published in the *Transactions of the Linnaean Society* (14:575) in 1825 using the type specimen collected by Mocino and Sesse in Mexico. Such is the tangled web of the scientific classification of these three rose shrubs.

In order to appreciate how the current accepted scientific names came into being, with *Cowania* dropped and *Purshia glandulosa* reduced to a variety, it is necessary to examine the natural ranges of the three shrubs. The range of antelope bitterbrush is generally described as British Columbia to Montana and south to New Mexico and California. Using the records of the Forest Service, USDA, August Hormay estimated that antelope bitterbrush occurred over 340 million acres of western rangeland in 11 states and the province of British Columbia, and on 7.5 million acres in California alone (Map 1.1).[5] Antelope bitterbrush occurs primarily in the great expanse of temperate desert environments between the Cascade–Sierra Nevada and Rocky Mountains. Many early authors took pains to point out that antelope bitterbrush did not occur extensively on the Pacific slope, being restricted to the Intermountain area. In California antelope bitterbrush has been collected in Trinity and western Siskiyou Counties, both of which are well west of the Cascade Mountains.

Hormay did not include the distribution of desert bitterbrush in his estimates. Desert bitterbrush is found in the mountains of southern Nevada, southwestern Utah, northwestern Arizona, extensively in the desert mountains of southern California, and extends into the mountains of Baja California, Mexico (Map 1.2).[6]

Map 1.1. General distribution of antelope bitterbrush in California and in the United States and southern British Columbia. Adapted from August L. Hormay, "Bitterbrush in California" (Res. Note 34, USDA, Forest Serv., Berkeley, Calif., 1943).

Cliffrose overlaps with desert bitterbrush in southern Nevada and southeastern transmountain California, and extends much farther eastward to include most of Utah, southwestern Colorado, western New Mexico, and all but southwestern Arizona. Its distribution also extends into northwestern Mexico.[7]

P. mexicana var. *stansburiana* is found only on the slopes of the Sierra Madre Occidental in south-central Mexico. MacArthur et al. (1983) considered *Cowania stansburiana* a separate species, although it has long been treated as a variety of *C. mexicana.* It grows from northern Mexico northward through California, Arizona, New Mexico, Colorado, Utah, and Nevada to about the forty-second parallel.[8] *C. stansburiana* generally has more leaf lobes than *C. mexicana,* a less revolute margin, less glandular formality on the leaves,

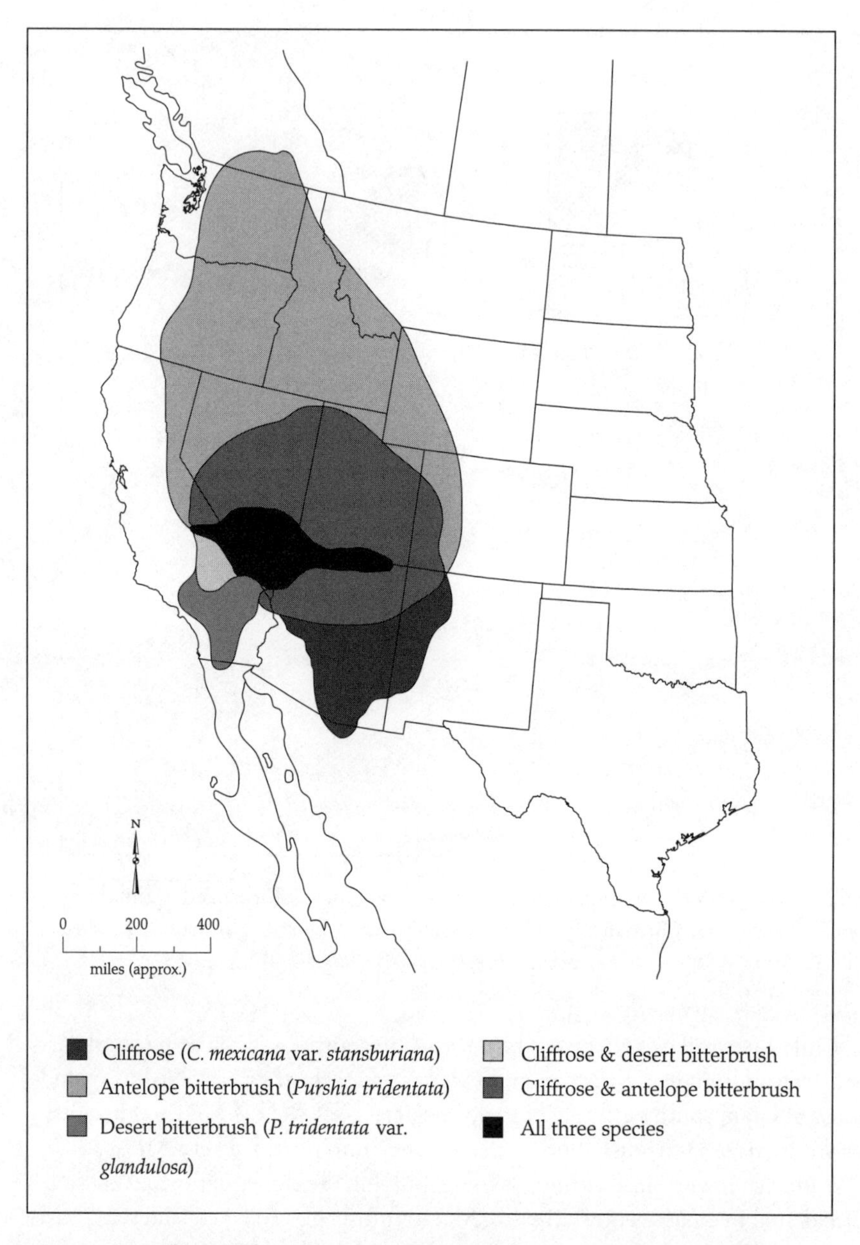

Map 1.2. Distribution of antelope and desert bitterbrush and cliffrose in western North America. Adapted from E. D. McArthur, H. C. Stutz, and S. C. Sanderson, "Taxonomy, Distribution, and Cytogenetics of *Purshia, Cowania,* and *Fallugia* (Rosoideae, Rosaceae)," *in Proceedings of the Symposium on Research and Management of Bitterbrush and Cliffrose in Western North America,* ed. A. R. Tiedemann and K. L. Johnson, 4–24 (Gen. Tech. Rep. 152, USDA, Forest Serv., Ogden, Utah, 1983).

a longer pedicel (the stalk of a single flower), more pedicel glands, larger sepals, a more funnelform hypanthium (the cup-shaped enlargement of the receptacle on which the calyx and corolla are inserted) angle, more stalked glands on flowers, and fewer achenes per hypanthium. *C. mexicana* has a pubescent hypanthium.

MacArthur et al. (1983) recognized three other distinct species of *Cowania*. *Cowania ericifolia* has a restricted distribution in limestone along the Rio Grande in the Big Bend area of Texas. Its reduced, needlelike leaves are an adaptation to drought. *Cowania subintegra* is taller than the straggling *C. ericifolia* but otherwise is very similar except that the leaves have irregular and infrequent side lobes. It has been collected only on calcareous shale near Bylas in southeastern Arizona. *Cowania plicata* is a very distinctive species with crimson pink flowers and much broader leaves than the other cliffrose species. It is native to the Sierra Madre Oriental in northeastern Mexico and occupies more mesic environments than the other *Cowania* species.

How, in terms of geologic times scales and plant evolution, did antelope bitterbrush, desert bitterbrush, and cliffrose come to occupy the landscapes where they are found today? The family Rosaceae consists of about 110 genera and some 3,000 species.[9] Its members occur worldwide but are more common in northern temperate regions, especially in western North America and eastern Asia.[10] The family Rosaceae is one of 17 families in the order Rosales, which probably derived from the primitive Magnoliales fairly early in angiosperm history.[11] Obviously, the evidence of the origins of the rose family is composed of incomplete bits and pieces obtained from the fossil plant and pollen records. The Rosaceae were well represented by the time of the Paleogene some 50 million years ago.[12] The order Rosales was associated with the northern group of continents (Laurasia) when Gondwana began to break into roughly the continents we know today.[13]

The three shrubs of interest to us fall into the subfamily Rosideae and the tribe Dryadeae. The common haploid (x) chromosome number of members of this subfamily is 7 or 9, with $x = 8$ or 14 also occurring. Polyploidy has been reported in this subfamily. Representative genera of the temperate deserts of the American West and their haploid chromosome numbers are *Purshia* (9), *Cowania* (9), *Cercocarpus* (9), and *Fallugia* (14).[14] The mountain mahogany species (*Cercocarpus*) occur sympatrically with the two bitterbrushes and cliffrose. Apache plume (*Fallugia paradoxa*), the only member of its genus, occurs in the Southwest from California to west Texas, northward to Utah and Colorado, and well south into Mexico (Map 1.3), generally in environments transitory from pinyon-juniper woodlands to more arid situations. Blackbrush (*Coleogyne* [haploid chromosome number = 8]) is found in a broad band across the

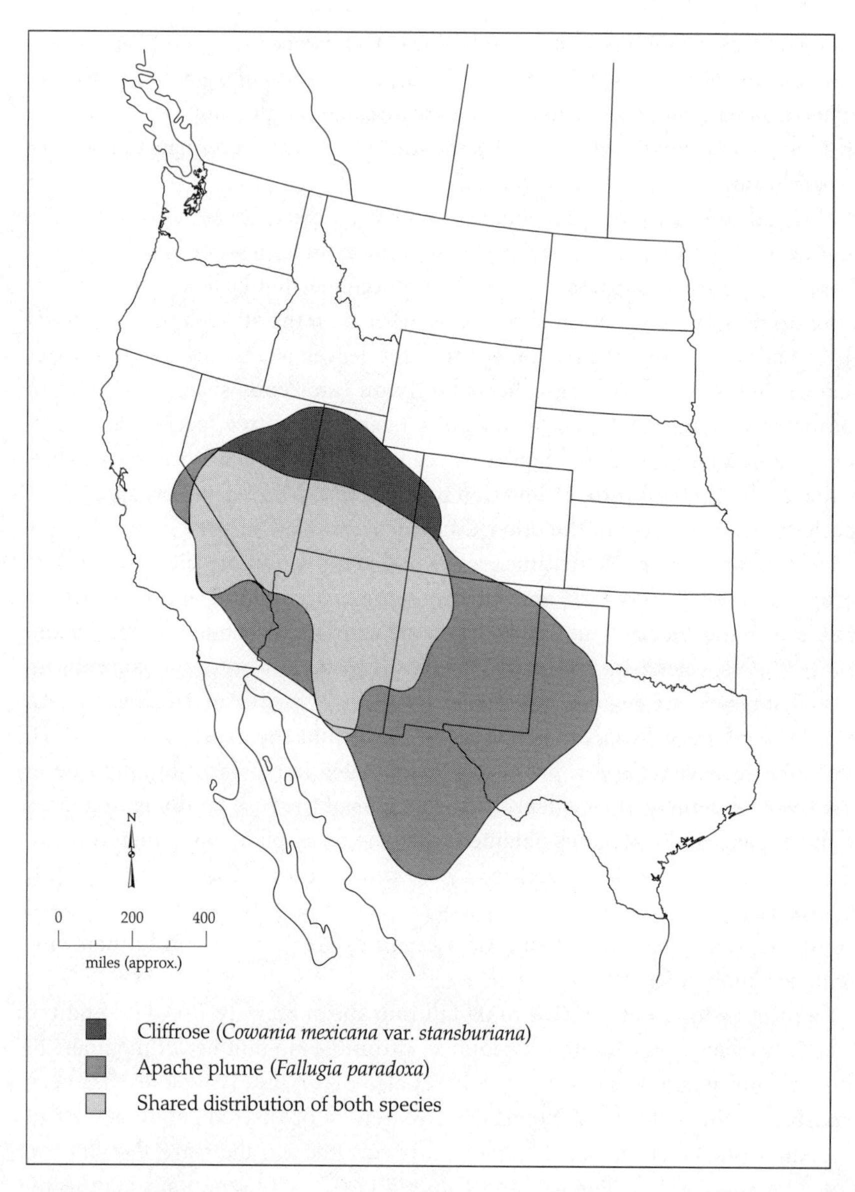

Map 1.3. Distribution of cliffrose and Apache plume in western North America. Adapted from McArthur et al., "Taxonomy, Distribution, and Cytogenetics of *Purshia, Cowania,* and *Fallugia,*" 1983.

southern Great Basin in areas in transition from sagebrush to warm desert ecosystems.

The more recent evolution of the geoflora of western North America can be explained by R. W. Chaney's concept of a northern Arcto-Tertiary geoflora and southern Neotropical Tertiary geoflora that came together in western North America. They in turn formed the basis of the Madro-Tertiary geoflora that arose in response to the development of seasonal, semiarid to arid environments.[15] D. I. Axelrod developed Chaney's concept further after examining leaf and fruit fossils from hundreds of locations in western North America, comparing them with museum specimens and extant plants, and determining the geologic age of the rock strata where the fossils were found.

Anyone with an interest in Great Basin plants should read Axelrod's description and analysis of Great Basin geofloras (*Mio-Pliocene Floras of West-Central Nevada*) and then visit the field sites.[16] The Eastgate Basin, located on U.S. Highway 50 in central Nevada, provides an excellent vista for comparison. If you arrive at the fossil site in late May in the early morning, you probably will become acquainted with one or more Great Basin rattlesnakes (*Crotalus viridis* subsp. *lutosus*). The site is in the salt desert with a sparse cover of shadscale (*Atriplex confertifolia*) and desert needlegrass (*Achnatherum speciosum*). The salt desert shrub communities grade into big sagebrush communities on the steeper alluvial fans that spill from the surrounding mountain escarpments down into the basin. On the rocky slopes above the alluvial fans a belt of pinyon-juniper woodland discontinuously clothes the Desatoya Mountains. Above these woodlands the mountain peaks support a mixed mountain brush community. Antelope bitterbrush occurs in the upper sagebrush, pinyon-juniper, and mountain brush communities.

D. I. Axelrod described the same vista as it probably appeared on a May day during the Miocene 10–15 million years ago. In his description, the template of mountains and basin is roughly similar. There is sufficient precipitation that a lake exists in the valley floor. Conditions for preservation of plant material are excellent because volcanic activity is building mountains to the west and a light rain of volcanic tephra is occurring. The foothills are open grasslands with occasional patches of sage (*Salvia*). On certain soils diverse species of oaks (*Quercus*) form woodlands. Higher on the mountain slopes chaparral patches occur where fires have burned in oak and mixed conifer woodlands. The upper slopes are dark with pine, fir (*Abies*), spruce (*Picea*), and Douglas fir (*Pseudotsuga*) trees. At higher elevations on north-facing slopes, stands of giant inland redwoods (*Sequoiadendron*) tower above the landscape. The redwood stands are greatly diminished compared with their extent at the start of the Tertiary, but logs from wind-thrown trees still float down into the shallow lakes, where minerals replace the wood in the petrification process. Add in a few broadleaf trees now found only in the southeastern United States or eastern Asia, and the Miocene landscape is complete.

If you cross the dry wash at Eastgate and walk among the eroding badlands of old lake sediments you are almost certain to happen upon an angular fragment of fossil bone. Obviously there were animals in Axelrod's Miocene Eastgate landscape. There is something about the texture and color of fossil bones that is eye-catching once you have found one. Perhaps the oddly shaped stone pseudomorph you hold in your hand was once part of an *Aepycamelus,* a giraffelike camel that browsed on trees in the woodlands. If you are lucky, you may find the fossilized tooth of *Neohipparion,* a horse that evolved in western North America. The mid-Tertiary vegetation of the Great Basin that was evolving under increasingly arid conditions was subject to selection pressure from numerous native large herbivores and browsers. The rose family species that were an intricate part of these plant communities responded to this selection pressure by adapting to the increasing aridity.[17] Because interactions with small granivorous (seed-eating) mammals became an important part of the seedling recruitment ecology of many of the rose family shrubs in semiarid environments, the symbiotic relations of plant and animals must already have been present in these mid-Tertiary forests and woodlands.

The Arcto-Tertiary flora was composed of a mixture of coniferous trees that we commonly associate with higher latitudes and altitudes, and deciduous temperate trees. The woody sagebrush species would represent a typical understory species of this Arcto-Tertiary geoflora. The Neotropical-Tertiary geoflora was dominated by broad-leaved evergreen species. In between these two geofloras, centered in the southern Rocky Mountains and adjacent Mexico, the sclerophyllous (thickened, hard leaves) and microphyllous (small leaves) Madro-Tertiary geoflora emerged. The Madro-Tertiary geoflora drew its constituent species from the stock of the other floras, but of necessity these plants became adapted to drier habitats.

The Madro-Tertiary geoflora is particularly rich in mountain mahogany and *Vauquelinia.* Fossil records for bitterbrush and cliffrose are spottier. Bitterbrush is recorded from the Pliocene some 10 million years ago.[18] Several rose family genera of the subfamilies Rosideae and Spiraeoideae have Madro-Tertiary affinities and the basic haploid chromosome number of 9. Besides sharing what is considered the primitive number of chromosomes (haploid = 9), these genera have similar growth habits and sclerophyllous or microphyllous leaves, and have species endemic to western North America. All these characteristics suggest that the Madro-Tertiary geoflora evolved in scattered pockets of functional aridity that may have been induced by rain shadows in the lee of growing mountain ranges or soil situations that did not favor moisture retention.

In his "Flora of the Providence Mountains" (1903) Brandegee described *Cowania mexicana* var. *dubia*, a form of cliffrose that differed significantly in appearance from *Cowania mexicana*. He suggested that this plant was a natural hybrid between cliffrose and bitterbrush and noted that it was similar to material collected in 1898 by J. A. Purpus (Joseph Anton Purpus, 1860–1932, botanical collector in the southwestern United States and Mexico) on Morey Peak in south-central Nevada.[19] Morey Bench is a noted deer winter range famous for its browse resources. Soon extensive hybrid swarms were discovered along the southeastern slopes of the Sierra Nevada and in an arc across the Great Basin where the range of bitterbrush and cliffrose overlapped. In a famous paper he published in 1959, the noted American geneticist G. Ledyard Stebbins used the *Purshia-Cowania* crosses as an example illustrating the role of hybridization in evolution.[20] Specifically, Stebbins used the desert shrubs to illustrate the principle of stabilization by introgression, a procedure, first suggested by Edgar Anderson in 1949, in which the products of hybridization become fixed and thus contribute to evolutionary change.[21] Three phases are essential to this process: the initial formation of F_1 hybrids, their backcrossing to one or the other parental species, and natural selection of certain recombinant types. Although this result of hybridization is to be expected whenever two populations with different adaptive norms hybridize, it is likely to be the commonest outcome if parental species are separated by well-developed barriers of reproductive isolation and are cross-fertilized. Under such circumstances, the relatively uncommon F_1 hybrids are much more likely to mate with members of their parental species than with each other, and the backcross individuals derived from such matings are likewise more apt to be viable and fertile than are progeny from $F_1 \times F_1$ matings. Stebbins suggested that because of the complexity of genetic mechanisms for adaptation possessed by species of higher plants and animals, new adaptations are more likely to develop by modifying the old ones, as happens with introgression, than by developing completely new gene combinations.

The products of introgression are permanent additions to the gene pool of the recurrent species. This is an inevitable result of the particular nature of heredity and the action of selection, which forms the third phase of the introgression process. When hybridization, recombination, and backcrossing result in the incorporation of new genetic material into the germ plasm of one of the original parents and it persists in subsequent generations, that material must impart a selective advantage. Under such circumstances, the relative permanence of genes, the homozygosity of the introgressive type, and the action of natural selection would combine not only to retain the introgressive type, but also to foster its spread throughout the area occupied by the habitat to which it is adapted. Stebbins concluded that introgressive genotypes persist indefinitely, can migrate far beyond the areas in which they originated, and can survive after the nonintrogressive parental species have become extinct.[22]

Howard Stutz and L. K. Thomas determined that populations of *Purshia* in British Columbia contain a few plants with bare, stalked hypanthium glands, suggesting some influence of *Cowania. Cowania* occurs far to the south of British Columbia, but apparently through the process of introgression *Cowania* germ plasm has spread to the northern extreme of antelope bitterbrush distribution.[23] We will develop the subject of the relative preference of browsing animals for antelope bitterbrush and cliffrose later. It is interesting to note here, however, that Stutz and Thomas suggested that introgression of *Cowania* into *Purshia* populations may be expressed in terms of animals' preference for the browse the different shrubs produce. Certainly such differences are evident in browsers' preferences for antelope bitterbrush.

McArthur et al. made a good case for cliffrose being the most ancient member of the desert rose shrubs.[24] Four factors point to *Cowania* as the ancestral species: (1) its distribution in the area where the Madro-Tertiary geoflora is believed to have arisen; (2) its greater intraspecific differentiation; (3) its primitive traits such as numerous flower parts and evergreen habit; and (4) its apparent isoenzymatic and morphological kinship with *Cercocarpus,* which is widely represented in the fossil record. *Cowania* appears to be one of the desert shrubs that Axelrod proposed was derived from Arcto-Tertiary and Neotropical geofloras in the Sierra Madre region of Mexico during the Tertiary.

Purshia is probably an early derivative of *Cowania* that subsequently evolved in isolation. The present contact between *Purshia* and *Cowania* is apparently of very recent origin. Despite the aridity of the Great Basin, dramatic changes in the distribution of major native plant species have been occurring during the Holocene. In the western Great Basin the pinyon has been spreading northward. Analysis of fossil packrat middens indicates that it reached its present northern limit of the Truckee River only within the last 1,000 years.

Antelope bitterbrush and cliffrose overlap in a zone that includes almost all of Utah, a large part of central and southern Nevada, and southeastern California. The only semblance of a reproductive isolation barrier is their partially disjunct flowering periods. As previously mentioned, antelope bitterbrush generally flowers earlier than cliffrose, but cliffrose plants growing on south-facing slopes often flower at the same time as antelope bitterbrush growing on north-facing slopes.[25] In such circumstances hybrids are abundant, mostly along ridge tops separating the parental populations. The hybrids are highly fertile and often backcross with the parents, providing an excellent example of the genetic principle of introgression. Introgression, or backcrossing, of antelope bitterbrush with cliffrose is less common than cliffrose providing the pollen to cross with antelope bitterbrush.[26]

The wholesale hybridization and subsequent introgression of antelope bitterbrush and cliffrose undoubtedly provide numerous new adaptive offspring. Sometime in the recent past this process produced desert bitterbrush. So recent is its origin that considerable genetic diversity can be demonstrated among geographically distinct populations. Despite this inherent variability, however, there is a suite of characteristics that consistently identifies this taxonomic unit as desert bitterbrush. The same can be said of all three of the species we discuss in this book: antelope and desert bitterbrush and cliffrose. It is obvious that they are related, and it is equally obvious that they can hybridize, and this fact has been used to justify reduction in the taxonomic classification units for the group.[27] On the other hand, they differ in many attributes used in taxonomy, including leaf size and shape, pubescence, glandulosity, time of flowering, number of stamens, number of carpels, deciduosity, and geographical distribution. The leaves of *Purshia, Cowania,* and *Fallugia* provide a fascinating array of similarity and divergence. Many other rosaceous genera (e.g., *Cydonia, Malus, Pyrus,* and *Sorbus*) have also demonstrated intergeneric fertility but are nevertheless maintained as distinct genera.[28] The practical natural resource manager may view the reduction of desert bitterbrush from a separate species to a variety or the change of cliffrose from *Cowania* to *Purshia* as mere taxonomic gamesmanship rather than good phylogenetic science. Within each of these classification units there exists significant variation in many characteristics that are vital for seed ecology, seedling establishment, and browse production and utilization.[29] The practical natural resource manager needs a finer rather than a more general classification system.

The fine-tuning of the classification system required to fulfill the needs of natural resource managers is probably not necessary at the species level, although few experienced field personnel would question the merit of separating antelope and desert bitterbrush into species. The recognization of the antelope bitterbrush cultivar Lassen illustrates both the utility and the problems with such finely tuned classification.[30] No one would deny that Lassen is a distinctive ecotype of antelope bitterbrush that has several characteristics of great significance to natural resource management. However, the outcrossing nature of antelope bitterbrush makes it difficult to define precisely the cultivar Lassen. In agronomic agriculture, in which cultivars are developed through the process of hybridization and selection with largely self-pollinated crops, the definition of specific cultivars is genetically very precise. It is, however, a trap beset with many scientific and practical perils to define Lassen antelope bitterbrush by the geographical area where it is native when continuous introgression occurs with adjacent populations. Users of wildland plants are going to have to develop a better system of nomenclature for scientific, practical, and commercial communication.

Chapter Two

Hunters, Herdsmen, and Brush

It took a long time for brush and the browse it produces to be recognized as an important component of rangeland production. Even after the birth of scientific range management early in the twentieth century, grass was looked on as the basic component of rangeland forage. Arthur W. Sampson, often considered one of the founders of scientific range management, was among the first to describe and discuss native range shrubs as components of the basic forage supply on ranges. Sampson's *Native American Forage Plants* (1924) is an excellent starting place from which to gain a perspective on the value of browse species and historical perceptions of shrubs in that role.[1]

In his discussion of browse plants Sampson lumped the rose family with the buttercups, willows, and miscellaneous other families. He recognized only two species of *Purshia,* both confined to the far western states, but concluded that the genus stood "high in the quality of its browse." Apparently, desert bitterbrush was the other species Sampson mentioned. He gave the general range for bitterbrush as from British Columbia to Montana, south to New Mexico, and west to the coast. "West to the coast" is interesting, because later authors limited the range to east of the crest of the Cascade–Sierra Nevada. Sampson described bitterbrush as a foothill to low-mountain plant growing up to an elevation of 8,000 feet. The most exceptional aspect of bitterbrush, Sampson said, is the fact that it flourishes where annual precipitation does not exceed 14 inches. This is an interesting way of expressing the plant's moisture requirements, because it gives no idea of the aridity that bitterbrush will actually tolerate. He indicated that bitterbrush produces luxuriant growth even on scabland soils. Just what he meant by "scabland soils" is not definitely known, but it might be a reference to the basalt soils of eastern Oregon and Washington. Sampson's experience in the eastern Great Basin is apparent in his association of bitterbrush with mountain mahogany, serviceberry (*Amelanchier*), and scrub oak (*Quercus*). He did not mention the pine/bitterbrush association of the Pacific Northwest and northeastern California. His eastern Great Basin prejudice is also apparent in his indication of the period of bloom for bitterbrush as extending from July 1 through August 10. In fact, over much of its range, the seeds of antelope bitterbrush are mature and dispersed by July 1.

Sampson noted that cattle, sheep, and goats closely browse bitterbrush throughout the growing season, and that sheep and goats particularly relish the browse of bitterbrush. Its very early bud burst and leaf growth are of particular value, he continued, because bitterbrush supplies a source of forage before herbaceous plants begin growth. This early turnout date would seem extreme under current range management practices.

Sampson also made note of the fact that bitterbrush was "said to be a 'strong feed' and to produce a solid fat which is not readily lost when animals browsed on it are shipped long distances or in inclement weather when feed is scarce." This statement reappeared in the literature (without the "it is said") for the next 50 years. Sampson concluded the section on bitterbrush with the comment that it had the additional asset of withstanding very heavy utilization.

Nowhere in his book did Sampson mention the dependence of mule deer on antelope bitterbrush for winter forage. In fact, deer (mule deer), elk, and wildlife are absent from the index for the entire volume, although later in the section on the rose family, in the subsection on true roses (*Rosa*), bitterbrush is described as an exceptional food source for wildlife.

We should not criticize Sampson too severely, for the literature available on bitterbrush when he prepared his book was quite limited. Bulletin 15 of the USDA, Bureau of Plant Industry, reported in 1902 on a range survey from the Columbia River to northwestern Nevada.[2] Bitterbrush was reported among the shrubs, including bitterbrush, fed on by sheep in the Warner Mountains of northeastern California and by numerous sheep bands in the mountains east of Jess Valley. Many shrubs were grazed down to short stumps, and herbaceous vegetation was entirely gone. The surveyors could not believe that the sheep were going to stay there the rest of the summer. This bulletin also contains perhaps the first photograph of bitterbrush ever published.

A study of sheep summer range in the Mica Mountains of eastern Washington published in 1913 lists bitterbrush as *Purshia tridentata* [Pursh] Spreng., an interesting choice of nomenclature because it admits that Pursh described the plant first.[3] (If the author had used Sprengler as an authority, the generic name *Kunzia* should have been used.) Although bitterbrush is listed, however, nothing is said about its importance on these rangelands.

Perhaps the first grazing experiment using bitterbrush was conducted on the eastern slope of the Pine Valley Mountains in the Dixie National Forest by C. L. Forsling and Earle V. Storm during the 1920s.[4] At the time of the study Forsling was the director of the Great Basin Experiment Station of the USDA, Forest Service, and Strom was a district forest ranger. Other shrubs surveyed in the study included red serviceberry (*Amelanchier rubescens*) and big sagebrush. The most preferred shrubs were mahogany (*Cercocarpus montanus*) and bitterbrush, which occurred largely in a decumbent form. It is not clear whether this growth form was natural for that site or had been induced by excessive utilization. The data collected were largely observational, but this was probably the first time detailed observations were made concerning seasonal use of bitterbrush. Grazing cattle used all of the current year's growth of the preferred shrubs; the heavily used shrubs were obviously declining in vigor or were already dead. The authors considered that 10–20 percent of each year's leader growth left unutilized would maintain vigor.

Under the common name buckbrush and the scientific name *Kunzia tridentata,* bitterbrush is mentioned in a 1918 bulletin reporting on the effect of grazing on yellow pine reproduction.[5] Apparently, the author attached little importance to bitterbrush as a browse species despite the fact that ranges were grazed so heavily that severe damage was occurring to pine reproduction.

In 1924, in a pioneering attempt to classify the natural vegetation of the American West, A. E. Aldous, classifier in charge of the Homestead Classification Unit, USDI, Geological Survey, and H. L. Shantz, physiologist in charge, Plant Geography, USDA, Bureau of Plant Industry, published "Types of Vegetation in the Semiarid Portions of the United States and Their Economic Significance."[6] The article lists antelope bitterbrush (*Purshia tridentata* [Pursh] DC.) as the dominant vegetation type along the Columbia River on sandy soils in Washington and Oregon and as also occurring in the Great Basin in association with sagebrush. In fact, bitterbrush was used as an indicator of soils not suitable for cultivation and crop production because of rock content. Bitterbrush reportedly provided fair carrying capacity for 20–30 head per section. Apparently, this refers to cattle carrying capacity, but it is not clear if it implies season-long grazing or spring and fall usage only. Most likely it means *all* forage and browse from range where bitterbrush occurred as a component of the rangeland community.

The concept of shrubs being an important part of the forage resources of western rangelands received vital stimulation in 1931 when William A. Dayton published *Important Western Browse Plants.*[7] In this USDA publication, Dayton, a plant ecologist working with the Forest Service, defined *browse* as the shoots or sprouts and leaves of woody shrubs, vines, and trees. The term *browse* was also applied to the process of cropping of this material by animals.

Dayton considered the rose family the most important browse-producing family on the western range because of its very wide distribution, although he pointed out that some genera in the family are seldom or never browsed. Dayton described the sites where bitterbrush grows as follows:

> Bitterbrush occurs on dry plains, hills, and mountainsides, mostly in well-drained, sandy, cindery, gravelly, or rocky soils, most commonly on south or southwestern slopes, up to about 3,500 feet elevation in the North and

> Northwest and 9,000 feet toward the southern end of its range. It is probably never found in typically wet or shaded situations and is frequently associated with species of sagebrush, snowberry, mountain-mahogany, serviceberry, and oak brush.

Concerning browse production, Dayton continued:

> Despite the characteristic taste of its herbage, alluded to in the common name "bitterbrush," or perhaps partly because of that fact, *Purshia* is one of the most important species of browse occurring on western rangeland [and] in some places is regarded as the most important single browse species in the locality. The abundant wedge-shaped 3-toothed leaves and the younger twigs, while seldom touched by horses, are extensively cropped by sheep, goats, and cattle, especially sheep. As the species is usually abundant, sometimes being the chief feature of the vegetational landscape, it is an important element of the carrying capacity. Its palatability appears ordinarily to be greatest in spring, winter and late fall, when the evergreen foliage and the usual large size of the plants enhance its utility. . . . In general, bitterbrush may be stated to have more value in southern Idaho, Utah, and the Southwest than in Oregon and the Northwest. In northern (especially northeastern) California, however, it is usually held to be good to excellent browse on sheep range. In many places in the west *Purshia* is one of the chief browse plants for game animals, being important as a winter and early spring feed for deer and antelope.

The illustration of antelope bitterbrush printed in this volume, by A. E. Hoyle, is perhaps the first ever published of the species (Fig. 2.1). Hoyle also contributed a beautiful drawing of Apache plume (Fig. 2.2).

Dayton's description of bitterbrush plants and habitat is remarkable both for what it says and for what it does not say. The mention of cindery soils suggests that the plant's occurrence in the vast trans–Cascade Mountain region of eastern Oregon and Washington with abundant pine/bitterbrush communities growing on volcanic tephra was at least recognized. Dayton stressed the abundance of bitterbrush, describing it as virtually the dominant shrub in many communities. Many of the earlier range studies, in contrast, list bitterbrush as a component species but fail to attribute special significance to it. What happened between 1900 and 1930 in terms of the perception of bitterbrush's importance? Did earlier authors simply not recognize bitterbrush as an important browse species, or was there a change in the plant's abundance during that 30-year period? Perhaps the abundance of bitterbrush dramatically increased during the first quarter of the twentieth century.

W. A. Dayton was in charge of range forage investigations for the Forest Ser-

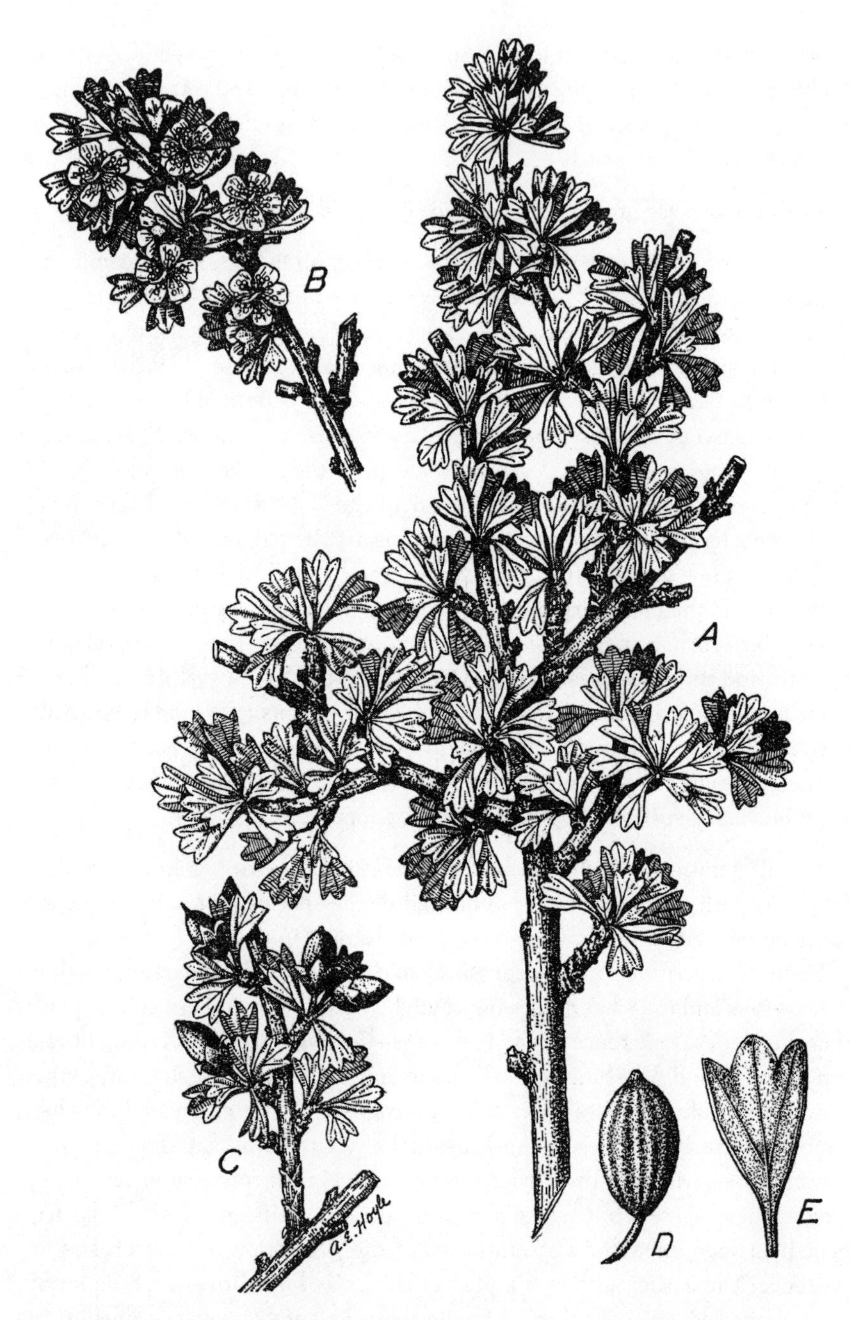

Fig. 2.1. Bitterbrush illustration from William Dayton's *Important Western Browse Plants* (1931). (a) sterile spray; (b) flowering spray; (c) fruiting spray; (d) fruit; and (e) leaf.

Fig. 2.2. Apache plume illustration from William Dayton's *Important Western Browse Plants* (1931). Cliffrose was not illustrated in this pioneering bulletin on browse species.

vice, USDA, when the agency published the *Range Plant Handbook* in 1937.[8] Among the listed contributors were Odell Julander, who became a noted mule deer habitat researcher, and Selar S. Hutchings, the first scientist to collect *Halogeton glomeratus*. The technical reviewers included W. R. Chapline, chief of range research, and Lincoln Ellison, then associate range examiner with the Northern Rocky Mountain Forest and Range Experiment Station of the Forest Service. Chapline had been clipping plots on the Jornada Experimental Range the day Pancho Villa raided across the border into Columbia, New Mexico. He remained professionally active into his mid-nineties. The purpose of the *Range Plant Handbook* was to evaluate 300 key plant species and to present succinct information on these species in an understandable, complete, and useful form. Each plant included was illustrated with line drawings (see Figs. 2.3 and 2.4).

The entry for bitterbrush has an introduction very similar to that previously used by Dayton's *Important Western Browse Plants*. It identifies bitterbrush as one of the most important browse plants on the western range, and on some ranges as *the* most important browse species. And for the first time the importance of bitterbrush growing as an understory species in ponderosa pine (*Pinus ponderosa*) woodlands is recognized. Pine/bitterbrush stands with Idaho fescue (*Festuca idahoensis*) understory vegetation are specifically mentioned for the Deschutes and Fremont National Forests of Oregon. Also singled out are the lava flats of northeastern California, where bitterbrush is reported to form extensive, dense, almost pure stands.

Dayton had previously indicated bitterbrush browse as being preferred in the winter and spring. The *Range Plant Handbook* indicates the shrub to be palatable at all seasons and to be preferred by all classes of domestic large animals except horses, although preferences are said to vary with location. The *Handbook* makes it clear that generally, throughout Utah, Colorado, Nevada, northeastern California, and in many parts of Idaho, *Purshia* was a prized forage species in the 1930s. In the Boise, Sawtooth, and Warner National Forests of Idaho its palatability ranged from worthless to only fair for sheep and poor to fair for cattle, even though on adjacent ranges it was regarded as an excellent browse plant. In eastern Oregon bitterbrush was rated as one of the most valuable browse species, while in adjacent Washington it was only moderately preferred, and in Montana it was considered only fair to good browse for cattle and sheep.

The variation in preference could be explained by the existence of alternative forage on given sites or by differences in plant chemistry due to the geological formations where the plants were found. Inherent differences in plant palatability were apparently not considered. The utilization reports used for the report were apparently all anecdotal accounts.

The *Range Plant Handbook* repeats A. W. Sampson's description of bitterbrush as producing "strong fat" that animals retain through adversity. Indica-

BITTERBRUSH

Pur'shia tridenta'ta, syn. *Kun'zia tridenta'ta*

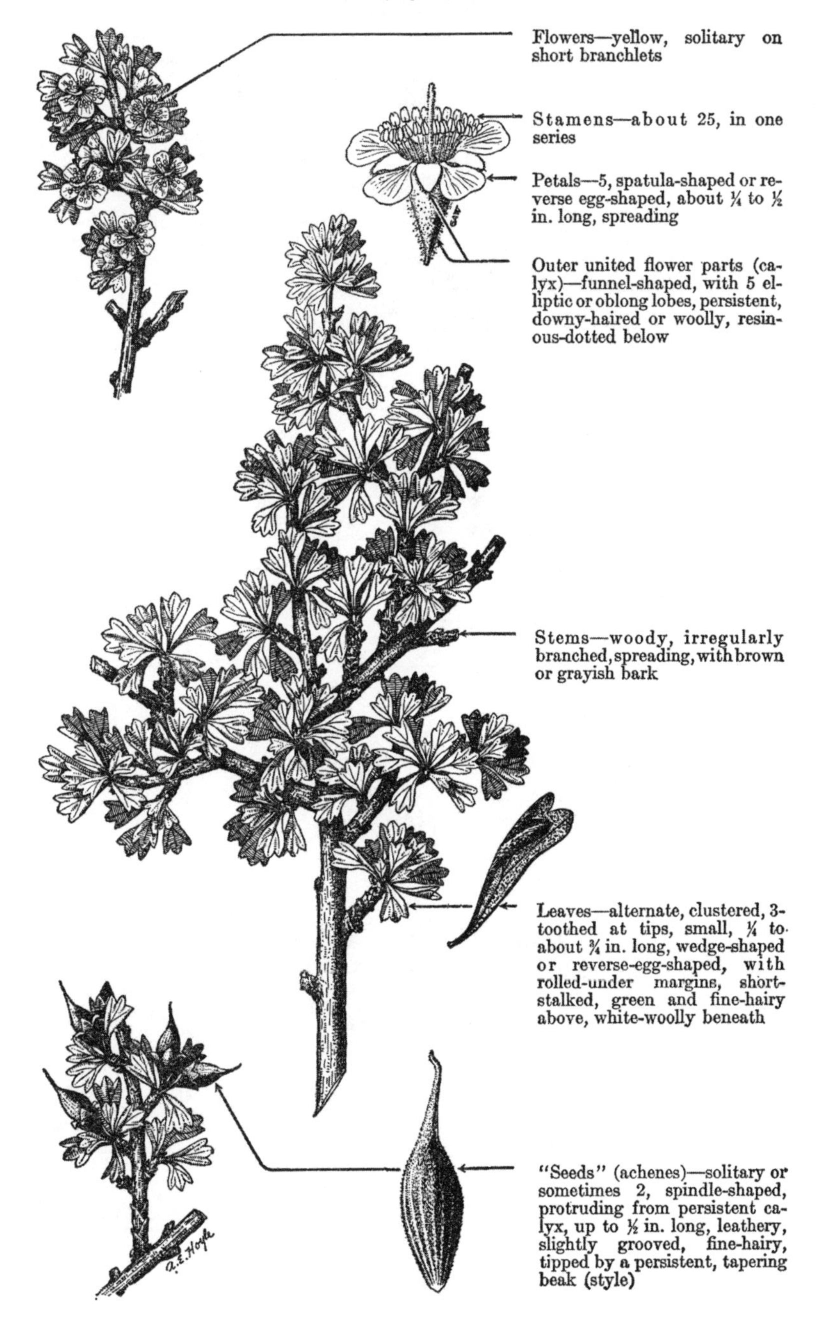

Fig. 2.3. Antelope bitterbrush illustration from the *Range Plant Handbook* (1937). Note that *Kunzia tridentata* is still listed as a synonym.

CLIFFROSE

Cowa'nia stansburia'na, syns. *C. al'ba, C. mexica'na stansburia'na, "C. mexica'na"* [1]

[1] Of United States authors, not D. Don.

Fig. 2.4. Cliffrose illustration from the *Range Plant Handbook* (1937).

tive of the high quality of bitterbrush as a browse is the account of high-quality fat lambs produced from dry ponderosa pine/bitterbrush ranges of the Deschutes National Forest. These lambs equaled or exceeded the quality of lambs produced on neighboring high-elevation mountain meadows in the Cascade Mountains to the west. The handbook also reports bitterbrush to be extremely resistant to overgrazing, although it notes that prolonged overgrazing can kill established plants or prevent reproduction.

The final paragraph of the *Range Plant Handbook*'s account of bitterbrush stresses the importance of the shrub on winter ranges for deer, elk, and antelope, citing Dixon's work in northern California as a source of information.[9] Actually, Dixon's data were limited to a few field observations of browsing mule deer in Lassen County, California.

Bitterbrush and Mule Deer

Starting about 1920, *bitterbrush* and *mule deer habitat* became synonymous terms. It is impossible to understand the environment where antelope bitterbrush occurs without first understanding mule deer. For much of the range of the bitterbrush species, mule deer are the most important native large herbivores.

American deer are descended from deer that crossed the Bering Land Bridge from Asia to North America early in the Pleistocene. This ancestral deer subsequently gave rise to the black-tailed, or mule, and white-tailed deer groups. Mule deer evolved in the mountains and semiarid to arid shrub lands of western North America, and white-tailed deer evolved largely in the subhumid to humid forests of eastern North America. The two species differ in external anatomy as well as in the obvious features of antler form, size and form of the tail, and size and form of the metatarsal gland.

With the onset of each glacial period during the Pleistocene the ancestral mule deer were driven south and downslope by the cold temperatures, and some persisted in the mountains of the southwestern United States and adjacent Mexico when the glaciers retreated. If this sounds familiar, recall that this is the same area where the Madro-Tertiary geoflora evolved. In the relative isolation of these mountainous islands several geographic races developed, and some of these expanded into the ranges released from the grip of Pleistocene ice. The northern limit of mule deer in Canada has been expanding ever since.

As with all classification systems, not all authors agree on the taxonomy of mule deer. We follow that presented by Cowan in his list of North American mule and black-tailed deer, which recognizes 11 subspecies:[10]

1. Rocky Mountain mule deer, *Odocoileus hemionus* subsp. *hemionus* Rafinesque
2. California mule deer, *Odocoileus hemionus* subsp. *californius* Caton
3. Southern mule deer, *Odocoileus hemionus* subsp. *fuliginatus* Cowan
4. Peninsula mule deer, *Odocoileus hemionus* subsp. *peninsulae* Lydekker
5. Inyo mule deer, *Odocoileus hemionus* subsp. *inyoensis* Cowan
6. Burro deer, *Odocoileus hemionus* subsp. *eremicus* Mearns
7. Tiburon Island deer, *Odocoileus hemionus* subsp. *sheldoni* Goldman
8. Desert mule deer, *Odocoileus hemionus* subsp. *crooki* Mearns
9. Cedros Island deer, *Odocoileus hemionus* subsp. *cerrosensis* Merriam
10. Columbia black-tailed deer, *Odocoileus hemionus* subsp. *columbianus* Richardson
11. Sitka deer, *Odocoileus hemionus* subsp. *sitkensis* Merriam

Rocky Mountain Mule Deer

This deer was first described in 1817 by Constantine Samuel Rafinesque (1783–1840)—the brilliant, eccentric pioneer Kentucky naturalist who was a prolific author of binomials for plants and animals—from fieldnotes made by Charles LeRaye as a captive of the Sioux Indians (1801–1812) on the Big Sioux River in South Dakota. The Rocky Mountain mule deer has the widest distribution of any subspecies of large game animal in North America. Its tolerance for a variety of climatic conditions has few parallels among large mammal species.

The western boundary of this subspecies' range extends generally along the crest of the Sierra Nevada and Cascade Mountains. In British Columbia, mule deer extend westward to the crest of the coastal mountains. In the early historic period, Rocky Mountain mule deer extended across the northern Great Plains to the woodlands along the Missouri River and down along the western Great Plains through Colorado. To the south the range reaches Trans-Pecos Texas and portions of the high plains in the Panhandle. Rocky Mountain mule deer are found in the mountains of New Mexico and west into the high country of the Mogollon Rim in Arizona.

California Mule Deer

The type specimen of this subspecies was collected in Gaviota Pass about 40 miles up the coast from Santa Barbara, California, by J. D. Caton in 1876. Its range includes coastal central California extending inland to the southern Sierra Nevada.

Southern Mule Deer

The type specimen was collected from the Barona Ranch east of San Diego, California. This race's range is confined to the mountains of extreme southern California and down the mountains of Baja California.

Peninsula Mule Deer

The type specimen was collected in 1896 between La Laguna and the Victoria Mountains in Baja California; known only from the southern part of Baja California.

Inyo Mule Deer

The type specimen was collected 10 miles west of Big Pine, California, at an altitude of 11,000 feet in 1911. This subspecies is distributed along the eastern slope of the southern Sierra Nevada, Inyo, and White Mountains.

Burro Deer

The type specimen was collected in the Sierra Seri near the Gulf of California opposite Tiburon Island, Sonora, Mexico, in 1895.

Tiburon Island Deer

The type specimen was collected from Tiburon Island, Mexico, in 1921; confined to Tiburon Island.

Desert Mule Deer

The type specimen was collected from the summit of the Dog Mountains in Hidalgo County, New Mexico. This subspecies is distributed in the deserts and mountains of extreme southern New Mexico, extending into adjacent Mexico.

Cedros Island Deer

The type specimen was collected from Cedros Island off the western coast of Lower California in 1896, and this race is found nowhere else.

Columbia Black-tailed Deer

The type specimen was collected in 1885 at Cape Disappointment in Pacific County, Washington, by hunters of the Lewis and Clark Expedition. The Columbia black-tail inhabits a narrow strip on the immediate Pacific Coast from Monterey County, California, to British Columbia.

Sitka Deer

The type specimen was collected in 1895 near Sitka, Alaska. The Sitka deer is restricted to the heavily timbered coastal strip from southeastern Alaska to British Columbia.

Several regional historical accounts describe the initial populations of mule deer that Europeans encountered, the decline of these herds following settlement and excessive harvesting through hunting, the regrowth of populations following protection, and the eventual decline of populations as the herds exceeded the carrying capacity of their habitats. In reviewing these historical accounts it is important to note how sparse the initial populations of mule deer apparently were in many areas, how rapidly the populations increased, and how huge the populations became before they crashed. It is surprising to discover how little was known about the diet of mule deer in the 1920s. In fact, J. S. Dixon appropriately titled a semipopular article published in 1928 "What Do Deer Eat?"[11]

In the Blue Mountains of northeastern Oregon and adjacent Washington, initial hunting destroyed elk (*Cervus canadensis* subsp. *nelsonii*) and Idaho white-tailed deer (*Odocoileus virginanus* subsp. *ochrourus*) populations and greatly reduced mule deer herds. Before long, the only remaining elk were found in the vicinity of Trout Meadows in the headwaters of the North Fork of the John Day River.[12] By 1900, hunting regulations and big game preserves had been established in the bistate area. Elk populations were enhanced by the reintroduction of animals from Rocky Mountain herds. Managers estimated that 30 percent of the winter range browse for both elk and mule deer was furnished by antelope bitterbrush.

Protection of the Blue Mountain mule deer herds resulted in tremendous increases by the 1920s. The Desolation Ranger District of the Whitman Forest, for example, had an estimated 360 elk and 3,100 mule deer in 1921. In 1931, E. P. Cliff estimated the elk and mule deer populations there to be 3,200 and 19,500, respectively.[13] The annual rate of increase, estimated at 23 percent for the mule deer population, was attributed to protection from hunting and reduced predator populations. Mountain lion (*Felis concolor*) and coyote (*Canis latrans*) populations were severely hunted and trapped at that time to reduce predation on domestic sheep. Cliff did not mention any change in shrub populations resulting from excessive domestic livestock grazing, but the introduction of large numbers of domestic livestock immediately preceded the initial decimation of the resident mule deer and elk populations. As perennial grasses were removed by continuous and improperly timed grazing, the shrubs became established and became productive browse producers.

Cliff estimated that the capacity of the Blue Mountain ranges to support elk and mule deer had been exceeded by 1929. Many western juniper (*Juniperus occidentalis*) and mahogany (*Cercocarpus ledifolius*) trees were highlined as high as browsing animals could reach, although others were untouched. Considerable mortality had occurred in antelope bitterbrush stands, and no seedling recruitment was observed. During the hard winter of 1931–1932 considerable mortality of mule deer occurred despite attempts to provide hay to isolated populations. Based on a sample of carcasses counted in April 1932, as many as 10,000 mule deer died during that winter. A series of mild winters after 1932 coincided with an increase in mule deer populations of about 15 percent per year.

C. M. Aldous, a biologist with the U.S. Fish and Wildlife Service (USDI), reported that before 1929 mule deer were not considered abundant anywhere in the Intermountain area, but afterward their population numbers exploded.[14] The famous Kaibab herd in northern Arizona, the Beaver Mountain herd in Fish Lake National Forest, and the Middle Fork area herd of the Salmon River in Idaho were noted examples of such population increases. He suggested four reasons for the increase:

1. Buck-only hunting laws were being enforced.
2. Many refuges for mule deer had been established.
3. Government agencies had been acting to reduce populations of predators.
4. The Forest Service had reduced livestock grazing on many national forest rangelands.

He did not initially associate the increase in shrub density and distribution caused by overgrazing with the increase in mule deer.

Aldous later conducted a study of mule deer herds in the White Pine County area of eastern Nevada, which had witnessed high winter mortality in the late 1940s. As part of the study he conducted utilization studies on key winter range browse species and determined that utilization of bitterbrush averaged 51 percent of current annual growth, cliffrose 38 percent, and mountain mahogany 100 percent. This study was one of the first to report crude fat, protein, and fiber content for key browse species. Aldous concluded that there were more mule deer than the winter range could support and that increased harvest of bucks and some does was necessary.

Oliver T. Edwards reported very similar results from a survey of mule deer habitat on national forest land in eastern Oregon.[15] Since 1932 there had been high winter mule deer mortality on the Murders Creek ranges of the Malheur National Forest. Bitterbrush was identified as composing about 33 percent of the mule deer winter diet there, and Edwards determined that it was being seriously damaged by overutilization. In a follow-up to this study, A. S. Einarsen reported on mule deer exclosures (which also excluded cattle and sheep) established in the Murders Creek area. Initially there was an increase in cheatgrass (*Bromus tectorum*), an exotic invasive species, in the exclosures. After several years, however, he noted an increase in the vigor of the native perennial grass Idaho fescue (*Festuca idahoensis*) and flowering of antelope bitterbrush plants.[16] Bitterbrush plants outside the exclosure did not flower. Einarsen, who was with the Cooperative Wildlife Research Unit at Oregon State University, seemed to be implying that the crash in the Murders Creek mule deer herd was entirely due to competition from domestic livestock. He did not mention the growth of the herd earlier in the century or the habitat conditions that made this growth possible.

A. Starker Leopold voiced another reason for the population growth and crash. Leopold was outspoken in relating the growth in deer populations to environmental changes caused by human activity.[17] In a famous paper presented at one of the North American Wildlife Conferences, Leopold related how the nineteenth-century mountain man John Work and his party of Hudson's Bay Company trappers ate some of their horses while crossing the Pitt River Valley of northeastern California in 1832 because they could not find game to kill. Yet this same area became famous in the mid-twentieth century for its huge mule deer populations and stands of antelope bitterbrush. In 1827, when Jedediah Smith crossed the Great Basin into California, he suffered some lean and hungry weeks traversing areas that are now well stocked with mule deer. Leopold concluded that most deer ranges were created when pristine natural resources were logged, burned, and grazed. Often, such deer habitat was created at an exorbitant cost in natural resources. Mule deer production during the first half of the twentieth century was at the expense of nutrients accumulated for centuries in the soils of old-growth forests and rangelands. The key to maximum densities of mule deer, Leopold insisted, was disturbance that resulted in the dominance of *woody* secondary succession species. And not only did shrubs increase in density on sites formerly dominated by trees, he insisted, they also came to dominate sites formerly dominated by bunchgrass. "On the east slope of the Sierra Nevada," he noted, " . . . such forage plants as sagebrush . . . and bitterbrush . . . have invaded foothills formerly stocked largely with bunchgrass." He considered a similar conversion to have occurred in much of the Great Basin. Leopold cautioned that fire and grazing were a two-edged sword in regard to mule deer habitat. They created suitable habitat, but they also had the potential to destroy what they had been instrumental in creating.

When Father Escalante visited what is now Utah in 1776, he recorded in his journal that there were few deer present and that wildlife in general was quite scarce.[18] Settlement of the Territory of Deseret, now known as Utah, began in the Salt Lake Valley in 1847. Big game was sparse then, too, and the pioneers suffered acute food shortages through the first winter. The livestock industry had spread by 1880 to most of the available range in Utah. Stockmen at the time reported deer to be present but by no means abundant. The few deer there were intensely hunted by Indians for their hides.

Early descriptions of Utah report the foothills to support extensive areas of bunchgrass with limited shrubs and juniper. By 1900 this had changed. The native vegetation had been severely affected by the concentrations of domestic livestock. Severe soil erosion and damaging floods were occurring on many rangeland watersheds. From 1877 to 1906 some 7 million acres were placed in forest reserves (which later became national forests) to improve rangeland management.

Utah issued its first hunting licenses in 1907. The licenses only cost a dollar, but big game was so scarce that they were not worth the price. There were not sufficient mule deer to have a season until 1913, and even then only 600 animals were killed in the entire state. In an effort to build up its mule deer populations the state began buying large acreages of winter range (about 1 million acres). The Dixie Forest in southern Utah did not have enough mule deer to make hunting worthwhile until 1927.[19] By 1930 there were already signs that the mule deer population was exceeding the carrying potential of the ranges that formed its habitat.[20] In 1935 the statewide mule deer population was estimated at 87,000, and 11,275 bucks were harvested. The population had increased to more than 200,000 by 1942. The hard winter of 1948 made it vividly apparent that mule deer ranges were overstocked when starvation killed thousands of deer throughout the state. In 1951 there were an estimated 300,000 mule deer in Utah, and the harvest (bucks and does) exceeded 100,000 animals. In less than 40 years, that is, the mule deer harvest had increased by 166 times (from 600 to 100,000). When Reynolds published this material in 1960 he was far ahead of other range and wildlife managers in recognizing the parallel between the dynamics of deer populations and the shift from grass to woody plant dominance.

At the time, wildlife managers were likely to invoke predator control and the establishment of hunting seasons and refuges as the reasons for the tremendous increase in mule deer populations. One of the basic principles of animal ecology says that the abundance of prey controls the size of predator populations because of the differential reproductive potential of the two classes of animals. In the case of mule deer populations, the reverse was true—lack of predators allowed the herds to increase. On the other hand, during the period of the vast expansion in mule deer herds that some have attributed to predator control, man, the most effective mule deer predator of all, was greatly increasing hunting pressure on the expanding herds. In Utah in 1925 there were 5,675 big game hunters; in 1935 this number had increased to 23,000, making a 407 percent increase in predation pressure.[21]

W. M. Longhurst, A. Starker Leopold, and R. E. Dasmann (among the most respected names in wildlife management in the mid-twentieth-century United States) conducted a survey of California deer herds for the state's Fish and Game Department that was published in 1952.[22] In the section concerning the Great Basin populations (mule deer and antelope bitterbrush range), the authors stated that bitterbrush was unquestionably the most important browse species. In contrast to the rest of California, where deer ranges were largely created by logging and wildfires, the Great Basin deer ranges were created by excessive grazing from domestic livestock. These grazing-induced, brush-dominated ranges had made the great expansion in mule deer populations possible. Logging and wildfires in the Sierra Nevada had increased summer habitat for mule deer, but excessive grazing of sagebrush-bunchgrass ranges in the trans-Sierra foothills had created the essential winter ranges. "To be sure," the authors noted,

> continued over-grazing of the *invading* [italics added] browse plants by both livestock and deer has tended to reduce the density of the more desirable species, like bitterbrush, permitting a disproportionate increase in less palatable shrubs like big sagebrush and rabbitbrush (*Chrysothamnus*). Thus many (probably most) browse ranges in the Great Basin region carry fewer deer then they might with more conservative use. Nevertheless, it is well to remember that livestock played a part in creating these ranges, even though too many livestock, especially in combination with too many deer, subsequently may destroy the best elements in the brush stands.

The authors considered the most serious threat to the Great Basin mule deer ranges to be the invasive annual cheatgrass, which enhanced the chances of ignition and the rate of spread of wildfires.

Most authorities agree that at the time of European contact, much of the prime mule deer ranges present during the early and middle twentieth century did not exist. Early explorers' accounts of the scarcity of mule deer support this hypothesis. Herbaceous vegetation apparently was a more significant portion of the environment under precontact conditions than woody species; that is, there was a lack of browse species. The changes in plant communities that favored eruptions of mule deer populations are both subtle and overt, and are not easily understood.[23] The key to understanding this is to be found in the synecology, physiology, nutrition, fire ecology, and management of antelope bitterbrush.

Chapter Three

Bitterbrush Plant Communities

Bitterbrush plants, by their abundance, stature, and extent, characterize certain communities. During the last three decades it has become popular among certain wildland scientists and land managers to develop vegetation classification systems based on the polyclimax theory of plant community ecology, which considers the potential natural vegetation of a site. This theory, which grew out of the efforts of Rexford Daubenmire of Washington State University and his many students, classifies the basic vegetation and soil units that identify a specific type of community in units known as habitat types.[1] (Often the term *association* is used instead of *habitat type*, but this can result in confusion. *Association* was originally defined as the fundamental unit of phytosociology, being a plant community of certain floristic composition, uniform habitat conditions, and uniform physiognomy.)[2] Habitat types are identified by the dominant species in each layer, usually starting with the tallest woody species. For example, P. T. Tueller et al. identified the following *Purshia* communities in Nevada:[3]

Northern Desert Shrub Communities

Purshia tridentata–Artemisia tridentata/Agropyron spicatum
Purshia tridentata–Artemisia tridentata/Sitanion hystrix
Purshia tridentata–Artemisia tridentata/Poa secunda
Cowania stansburiana–Artemisia tridentata
Purshia glandulosa–Artemisia tridentata
Purshia glandulosa–Artemisia tridentata/Prunus andersonii

In the first example, antelope bitterbrush and big sagebrush share dominance in the shrub layer with an understory of bluebunch wheatgrass (*Agropyron spicatum* = *Pseudoroegneria spicata*). (Where the plant nomenclature has been changed since the community was published we present the modern nomenclature the first time the name is mentioned.) The cliffrose community is a two-shrub dominant community with no herbaceous dominant species specified.

Pinyon-Juniper Woodland

Juniperus osteosperma/Purshia tridentata
Juniperus osteosperma/Purshia glandulosa–Artemisia tridentata
Juniperus osteosperma/Cowania stansburiana–Artemisia tridentata

These juniper woodland communities comprise a tree overstory with bitterbrush or cliffrose in the understory.

The basic idea in the polyclimax classification system—to group like environments of the same potential—has great theoretical value in making natural resource management decisions. In practice, however, it has led to a list of described communities as long as a telephone book, and when you dial the number of a specific community, nobody answers. Too often, use of the system results in a series of abstractions created by the person developing or enlarging it. These abstractions reflect only portions of the natural and human-induced variation contained in the system. Sampling in the center of a *selected* (not randomly selected) site and drawing the extremes to the modal expression of the community tend to create discrete communities in what is actually incompletely represented continuous variability.

This discussion may seem unnecessarily complex, but the practical point is that there are actual landscapes that, because of the density, appearance, and spatial distribution of bitterbrush and associated species and the topographic position, are identified as "bitterbrush potential sites." The point of this chapter is to describe the extent, composition, and structure of such communities. Our most important goal in these community (or synecological) studies is to determine whether current bitterbrush communities exist as climax or near climax communities or are instead some form of a succession disclimax induced by human disturbance in the form of domestic livestock grazing, changes in fire frequency and intensity, or artificially manipulated native big game populations. Why is this important? If current big game populations depend on a bitterbrush resource that is *not* being renewed, those populations are going to crash when the stands become decadent.

Considering that the concept of habitat types originated in the Pacific Northwest, it is fitting that we begin our description of the kind of communities where *Purshia* species grow in that region of the country. Starting in the interior of British Columbia and flowing down through portions of the Columbia Basin and along the northern Rocky Mountains is a repetitive series of plant communities that contains antelope bitterbrush. These communities range from pine woodlands at higher elevations down through juniper woodlands in the foothills to intermingle in selected examples of shrub-steppe communities. The companion woody and herbaceous species change as this continuum of communities extends through portions of eastern Oregon and western Idaho to northern California and Nevada. The variability in the community composition and the types of sites where the communities occur is considerable. The basic sameness of the communities despite this variability is astonishing.

The classic summary of synecological information for the Pacific Northwest is *Natural Vegetation of Oregon and Washington*, by Jerry Franklin and C. T. Dyrness, which makes numerous references to communities that support *Purshia tridentata*.[4] In the western juniper (*Juniperus occidentalis*) zone the authors recognized five major communities. Antelope bitterbrush is present in four of these five communities. In three of the four, *Juniperus/Artemisia/Festuca*, *Juniperus/Artemisia/Agropyron*, and *Juniperus/Artemisia-Purshia* had 100 percent constancy (constancy refers to the number of stands sampled in which the species was present) in all the examples of these communities that had been sampled and reported in the literature. About one-third of the stands sampled in the fourth community, *Juniperus/Artemisia/Agropyron-Astragalus*, contained antelope bitterbrush. In all of these communities, the sagebrush is big sagebrush, *Artemisia tridentata*. At the time these studies were conducted, the practice of identifying big sagebrush to the subspecies level was not employed.

Much of the information about the juniper woodland communities was gathered by Richard Driscoll.[5] He described the soils of the *Juniperus/Artemisia-Purshia* communities as being deep (45 inches) sandy loams without horizon development over a buried soil. In the trans-Cascade region, repeated volcanic tephra falls have created many buried soils, and the roots of the larger shrubs extend into the buried soil. This is a shrub-dominated understory with 8.2 percent cover of big sagebrush (Table 3.1). Herbaceous perennials are few and widely spaced. The western juniper cover averaged only 6.6 percent, the lowest of any of the juniper communities Driscoll described. The photographs Driscoll used to illustrate his publication show the western junipers as vigorously growing pole-sized trees.

Driscoll found *Juniperus/Artemisia/Festuca* and *Juniperus/Festuca* communities on northwest- to northeast-facing slopes where there was greater soil development and more effective soil moisture for plant growth. Western juniper cover averaged about 12 percent. The *Lupinus* variant of this type of community tended to have the western juniper in clumps with much evidence of recent wildfire activity. Antelope bitterbrush was absent from the examples of this type that Driscoll sampled.

Antelope bitterbrush had 80 and 100 percent constancy, respectively, in the *Juniperus/Agropyron* and *Juniperus/Festuca* stands Driscoll sampled. The soils of these communities are strongly developed loams derived from basalt. Western juniper cover averaged 10 percent. One of the high-constancy perennial

TABLE 3.1.

Mean Percentage Foliage Cover and Constancy for a Juniperus/Artemisia-Purshia *Association in Eastern Oregon*

Species	Association	
	Cover	Constancy
Juniperus occidentalis	6.6	100
Artemisia tridentata	8.2	100
Festuca idahoensis	2.5	100
Agropyron spicatum	0.9	100
Poa secunda	0.2	60
Koeleria cristata	0.1	100
Achillea millefolium	0.2	100
Lomatium triternatum	—	20
Collinsia parviflora	0.2	100
Phlox douglasii	—	20
Gayophyium lasiospermum	1.0	100
Eriophyllum lanatum	—	40
Bromus tectorum	2.7	100
Stipa thurberiana	0.3	60
Astragalus sp.	—	60
Cryptantha ambigua	0.6	100
Chrysothamnus nauseous	0.2	40
Sitanion hystrix	0.2	100
Erigeron linearis	0.1	100
Purshia tridentata	5.5	100
Chrysothamnus viscidiflorus	1.5	100
Collomia grandiflora	0.1	100
Mentzelia ablicaulis	0.5	100
Montia perfoliata	0.1	100
Stipa comata	—	80
Linanthus harknessii	—	20
Eriogonum baileyi	—	20
Eriogonum umbellatum	—	20
Total perennial herb cover	4.6	
Total shrub cover	16.2	

Adapted from R. S. Driscoll, "Vegetation-Soils Units in the Central Oregon Juniper Zone" (Res. Paper 19, USDA, Forest Service, Portland, Ore., 1964).

grasses in these communities is Thurber's needlegrass (*Stipa thurberianum* [*Achnatherum thurberianum*]). In the central Great Basin, *Artemisia/Stipa thurberianum* communities become landscape dominants.

The western juniper woodlands of eastern Oregon extend southward into northeastern California with a few examples in northwestern Nevada.[6] The same types of communities are also represented in the Owyhee uplands of southwestern Idaho. The famous Devil's Garden mule deer herd of Modoc County, California, has most of its winter range in western juniper woodlands.

The western juniper woodlands extend westward from Modoc and Lassen Counties into Siskiyou and Trinity Counties. West of Mount Shasta, antelope bitterbrush is found growing on soils derived from sediments in woodlands dominated by Oregon oak (*Quercus garryana*). Antelope bitterbrush is found even in relatively pure communities on soils derived from serpentine. An isolated, disjunct population of antelope bitterbrush occurs in a western juniper woodland, growing on serpentine, in the Yolla Bolly Mountains of the north Coast Range.[7]

When most of the early western juniper synecology work was completed in the 1950s and early 1960s, ecologists did not generally recognize the serious consequences for mule deer of increases in the density and distribution of western juniper trees. The change in western juniper density and distribution, and similar changes in pinyon-juniper woodlands, was the most significant factor affecting mule deer habitat during the twentieth century. We develop this subject in a later chapter.

The relative abundance of *Juniperus occidentalis/Artemisia arbuscula* communities is apparently greater in northeastern California than in eastern Oregon. On the Modoc Plateau, the acreage of big and low sagebrush communities is about equal, while in the Humboldt River basin of northwestern Nevada, big sagebrush communities constitute 40 percent and dwarf sagebrush species characterize only 5 percent of the total landscape.[8]

For much of eastern Oregon, the Modoc Plateau of northeastern California, and the Owyhee Mountains of southwestern Idaho, the western juniper/big or low sagebrush–antelope bitterbrush/bunchgrass communities are remarkably similar in appearance. The repetitive basalt flow landscape with sudden rimrocks and occasional conical volcanic cones imparts uniformity to these landscapes. The gray-green foliage of western junipers once provided a visual enhancement in the overwhelming silver gray of sagebrush-flooded valleys. The old-growth juniper stands were tucked in against the blackness of the basalt rims and talus slopes. Within the twentieth century the juniper woodlands surged across the landscape, drowning sagebrush and antelope bitterbrush communities and replacing the silvery gray of sagebrush with dusky green.

Perhaps the most famous antelope bitterbrush communities occur from Su-

Fig. 3.1. Eastern slope of the far northern Sierra Nevada showing Jeffrey pine–ponderosa pine–California black oak woodlands with antelope bitterbrush understory. A narrow band of antelope bitterbrush–basin big sagebrush communities occurs at the base of the slope before the lake plain of Honey Lake Basin.

sanville to Doyle, California, along the eastern flanks of the extreme northern Sierra Nevada (Fig. 3.1). The density, the 10–14-foot height of these plants, and the proximity of the communities to U.S. Highway 395 make these stands hard to miss. This is the ecotype of antelope bitterbrush that has been described as the cultivar Lassen. These sites are not strictly western juniper woodlands, although they are sandwiched between western juniper woodlands to the north and south; rather, they are California black oak (*Quercus kellogii*) woodlands, at least on the Susanville end of the distribution (Fig. 3.2). These are the only black oak woodlands in the Great Basin, and this may be the only location along the eastern front of the combined Cascade–Sierra Nevada where juniper or pinyon-juniper woodlands do not form an interface between the interior steppe vegetation and the pine woodlands of the mountains. The soils of these Lassen bitterbrush sites are often described as deep sands derived from decomposing granite. The alluvial fans that support the stands of Lassen bitterbrush are actually mixtures of granite and recent basalts and andesites. The true California black oak woodlands form a narrow fringe below pine woodlands (which will be discussed later) that extend down from the steep northern Sierra Nevada slopes. The Lassen cultivar extends below the California black oak woodlands and well

Fig. 3.2. Landscape view from Janesville Grade, Lassen County, California, looking north across the Susan River Valley. In the foreground are Jeffrey pine–ponderosa pine–California black oak/antelope bitterbrush communities. Bass Hill, in the center, was once an important mule deer winter range.

below the maximum elevation (4,380 feet) of pluvial Lake Lahontan. A few relatively undisturbed stands of *Purshia tridentata–Artemisia tridentata/Stipa comata (Hesperostipa comata)–Oryzopsis hymenoides (Achnatherum hymenoides)* can still be found on sand-textured soils on the lake plain below Janesville. These stands offer a sample of how these communities might have appeared before extensive housing development and repeated wildfires greatly depleted their extent. The codominant sagebrush is basin big sagebrush.

When the leaves of the California black oak turn yellow-orange in the autumn, the up-slope evergreen pine woodlands provide the perfect backdrop to accentuate the colors. Residents of the eastern Great Basin expect the oak brush, maple, and aspen colors of the Wasatch Front, but in the West, even the black oak colors are a rare treat.

The most famous mule deer winter range within the home range of the Lassen cultivar is Bass Hill. Extending diagonally, partially across the Honey Lake Basin from the Sierra Nevada, this ridge supports dominant antelope bitterbrush–big sagebrush/blue bunch wheatgrass communities with occasional areas of low sagebrush–antelope bitterbrush mosaics on shallow soils. In contrast to the alluvial soils at the base of the adjacent Sierra Nevada, the Bass Ridge communities are located on residual soils derived from basalt-andesite. By the mid-1980s the mature stands of antelope bitterbrush were 15-plus feet in height and the perennial grass understories had been replaced by the invasive annual cheatgrass. The bulk of these stands was destroyed in a spectacular wildfire that featured 50-foot flames.

On the opposite side of the Honey Lake Valley from the Lassen bitterbrush sites are badly depleted communities of a dwarf form of antelope bitterbrush growing on vertisol clays in depressions on basalt flows. These communities contain Lahontan sagebrush (*Artemisia arbuscula* subsp. *longiloba*) and sage (*Salvia dori*). An early map of the distribution of bitterbrush species in California, completed by Eamor Nord, shows the Honey Lake Valley to be free of antelope bitterbrush.[9] On what is known as the island in the middle of the valley are extensive stands of tall antelope bitterbrush growing on sand dunes that are superimposed on pluvial lake sediments. This contrast in the stature, associated soils, and phytosociological association of antelope bitterbrush in one geographical area does much to underscore the breadth of this species. And it is important to remember that all of these antelope bitterbrushes can freely exchange genetic material through hybridization. The only barrier to this exchange is differing flowering periods.

The western juniper woodlands with associated antelope bitterbrush communities end just north of Reno, Nevada, and with a few small, isolated exceptions, the pinyon-juniper woodlands of the Great Basin begin south of the Truckee River, at Reno. These distribution statements need some qualification. The pinyon-juniper woodlands that occur at the Truckee River do not touch the eastern slope of the Sierra Nevada except for a small patch of Utah juniper (*Juniperus osteosperma*) in western Reno. Trees that morphologically and genetically appear to be western juniper occur as far south as the west end of the Sweetwater Mountains in Mono County, California. This stand contains antelope bitterbrush plants.[10] *Juniperus occidentalis/Artemisia tridentata–Purshia tridentata/Stipa occidentalis* (*Achnatherum occidentalis*) communities occur in Long Valley, north of Reno. *Juniperus occidentalis/Artemisia tridentata–Purshia tridentata/Agropyron spicatum* communities occur on the Seven Lakes and Tule Peak Ranges at the extreme southern range of western juniper. In the highlands of Plumas County, California, west of the Sierra crest, antelope bitterbrush occurs in scattered juniper stands. These juniper trees appear to be western juniper, but genetic testing indicates them to be highly influenced by Sierra juniper (*Juniperus australis*), which occur down the crest and on eastern slopes of the Sierra Nevada. In Little Antelope Valley, Mono County, California, Sierra juniper woodlands with antelope bitterbrush are found on the west side of the valley and pinyon–Utah juniper woodlands occur one-half mile away on the east side. Population genetics studies conducted by Robin Tausch indicate that the relative genetic impact of western juniper on Great Basin woodlands is much more complex than this simplistic division, however.[11]

Franklin and Dyrness reported an extensive area with landscapes characterized by ponderosa pine (*Pinus ponderosa*) above the western juniper woodlands of eastern Oregon and Washington. Community composition in ponderosa pine stands varies widely with geographic location, soils, elevation, and succession status. The history of stand disturbances (e.g., by fire and logging) influences the density of the overstory, which in turn can have profound effects on the composition and density of the understory.[12] The open nature of typical mature trans-Cascade ponderosa pine stands provides abundant niches for sun-loving species, including many typical of steppe and shrub-steppe communities.

Ponderosa pine communities have been described from south-central Oregon through the Blue Mountain province of northeastern Oregon to the mountains of eastern Washington. Neil West suggested that these woodland communities can be explained by analyzing continually variable environmental gradients.[13] Franklin and Dyrness chose to treat the province as a series of distinct communities, although they admitted that such species as *Purshia tridentata* and *Festuca idahoensis* occurred throughout much of the range of the type.

Ponderosa pine woodlands provide the visitor with a banquet of visual splendor. Old-growth stands with mottled golden brown boles in a spacious, parklike arrangement beneath a vaulted canopy of dark green crowns are the woodland treat of the Far West. Viewed in the slanting light of the early morning or evening, the trunks seem to emit light, and the depth of the stand appears infinite. The distinctive dark green of rigid antelope bitterbrush leaves on black stems enhances the appearance of these woodlands.

Among the six ponderosa pine associations recognized in eastern Washington is *Pinus ponderosa/Purshia tridentata.*[14] This association has a *Purshia*-dominated shrub layer superimposed on a variety of perennial grasses including *Festuca idahoensis, Agropyron spicatum, Stipa comata,* and *Aristida longiseta.* In some stands, forbs such as *Balsamorhiza sagitta* and *Erigeron compositus* are abundant. Antelope bitterbrush occurs in stands of many of the other ponderosa pine associations and extends into British Columbia.[15]

In the Blue Mountains, *Pinus ponderosa/Agropyron spicatum* and *Pinus ponderosa/Purshia tridentata/Agropyron spicatum* associations are often found in areas of transition between steppe and shrub-steppe and forests. In the extreme southern Blue Mountains, a *Pinus ponderosa/Purshia tridentata/Carex rossii* association is found on some coarse-textured soils.[16]

In south-central Oregon the soils are highly influenced by volcanic tephra from the Cascade Mountains to the west. Sclerophyllous shrubs such as *Arctostaphylos patula* and *Ceanothus velutinus* assume importance in these areas. Total plant cover tends to be lower, especially in the more xeric communities, and the herbaceous flora is more depauperate on pumice soils. Dyrness and Youngberg enumerated five ponderosa pine plant associations for this region,[17] four of which include significant amounts of antelope bitterbrush:

Pinus ponderosa/Purshia tridentata
Pinus ponderosa/Purshia tridentata/Festuca idahoensis
Pinus ponderosa/Purshia tridentata–Arctostaphylos patula
Pinus ponderosa/Ceanothus velutinus–Purshia tridentata

The authors considered these communities edaphic disclimaxes because of the immaturity of the pumice soils where they occur. The *Pinus ponderosa/Purshia* community is at the lowest elevation in the ponderosa pine woodlands and is characterized by open stands with minimum herbaceous vegetation. Characteristic species are *Stipa occidentalis, Sitanion* (*Elymus elymoides*), *Gayophytum,* and *Cryptantha affinis. Pinus/Purshia/Festuca* communities occur on finer-textured soils derived from water-laid pumice deposits. Characteristic species include *Stipa occidentalis, Carex rossi, Achilles millefolium, Paeonia brownii,* and *Eriophyllum lanatum.*

The *Pinus/Purshia-Arctostaphylos* communities are situated at slightly higher elevations than the *Pinus/Purshia* ones. This community shares many species with *Pinus/Purshia,* but also contains seral *Pinus contorta* and additional herbaceous species such as *Phacelia heterophylla.* In the *Pinus/Ceanothus-Purshia* communities, *Ceanothus velutinus* replaces *Arctostaphylos patula* in the shrub layer. More abundant tree reproduction and occasional patches of willow (*Salix*) indicate more mesic conditions.

Franklin and Dyrness mentioned a series of *Pinus ponderosa/Purshia tridentata/Carex pennsylvanica* communities with various associated shrub species. On residual soils in south-central Oregon that are not covered by recent pumice, the dominant pine woodland is *Pinus ponderosa/Purshia tridentata/Festuca idahoensis.*[18]

Fred Hall reported on nonforested communities in the Ochoo Mountains of eastern Oregon that occur in association with ponderosa pine woodlands.[19] These communities include *Artemisia arbuscula/Agropyron spicatum–Poa sandbergii* with *Purshia tridentata* as a major associated species. Low sagebrush (*Artemisia arbuscula*) in this mosaic of woodlands and shrub-dominated openings is associated with clay-textured surface soils that are saturated with moisture in the early spring. Hall also reported the occurrence of a *Purshia tridentata–Cercocarpus ledifolius* community that characterizes rimrock in the lower ponderosa pine zone.

The ponderosa pine/antelope bitterbrush–type community is perhaps the most extensive—and therefore the most diverse—of the *Purshia*-characterized landscapes. Extending east from the Wallowa Mountains, *Pinus ponderosa/Purshia tridentata* communities are found across the southern Idaho batholith, mostly on dry south-facing slopes on the lower timberline. There are two understory phases: the Idaho fescue typical of northern Idaho and eastern Oregon, and bluebunch wheatgrass.[20] Similar ponderosa pine communities occur in Rocky Mountain National Park in Colorado.[21] An associated woody species in these communities is *Cercocarpus montanus.*

We stated above that the important issue when discussing bitterbrush communities is to establish differences among climatic climax communities and various disclimax situations induced by soils, fire, grazing, or logging. J. M. Peek et al. reported on a ponderosa pine/antelope bitterbrush stand in Idaho that they considered to be actually a potential *Pseudotsuga menziesii/Symphoricarpus albus* community.[22] The current community had been created by periodic burning and maintained by mule deer and elk browsing.

The pine woodlands of Oregon extend down into northeastern California. The Medicine Lake Highlands, extending down to Mount Shasta and across the Pit River to the Mount Lassen Highlands, have many discontinuous examples of the previously described pine/antelope bitterbrush communities. The associated species depend on soils, aspect, and elevation.[23] A constant understory shrub in many of these pine/antelope bitterbrush woodlands is prostrate ceanothus (*Ceanothus prostratus*).

Farther south along the eastern flank of the Sierra Nevada, ponderosa pine is replaced by Jeffrey pine (*Pinus jefferyi*). Extensive Jeffrey pine woodlands with antelope bitterbrush understories occur in the eastern slope uplands as far south as Lake Tahoe. From Tahoe south, the eastern slope is precipitous enough virtually to eliminate pine woodlands of the broad, undulating upland type found north of the lake. Jeffrey pine–mountain mahogany–Sierra juniper–antelope bitterbrush stands occur sporadically along the eastern side of the southern Sierra Nevada. South of the Carson River, pinyon-juniper woodlands meet the woodlands of the Sierra.

The Janesville Grade area of Lassen County, California, where the native range of the Lassen cultivar of antelope bitterbrush occurs, also provides unusual examples of pine/antelope bitterbrush woodlands. The upper third of the alluvial fans at the eastern base of the farthest northward extension of the Sierra Nevada support ponderosa–Jeffrey pine–California black oak/antelope bitterbrush woodlands. The robust form of antelope bitterbrush and the black oak extend well up the mountain escarpment until true firs (*Abies*) mingle with the pine overstory. On the undulating uplands that occur west of the escarpment, antelope bitterbrush is common in pine–greenleaf manzanita (*Arcto-*

Fig. 3.3. Jeffrey pine/desert bitterbrush community south of Mono Lake.

staphylos patula)/Idaho fescue woodlands, but it is a low-growing form of *Purshia* typical of the pine woodlands of northeastern California.

The recent volcanic uplands south of Mono Lake provide habitat for extensive areas of Jeffrey pine–bitterbrush woodlands (Fig. 3.3). Eamor Nord considered these to be desert bitterbrush (*Purshia tridentata* var. *glandulosa*) plants. It is our observation that the transition from antelope bitterbrush to desert bitterbrush along the eastern side of the Sierra is far too complex to separate into distinct species. The pine/*Purshia* woodlands south of Mono Lake are some of the more striking examples of this type in western North America. The woodlands occur at high elevations on stark, recent volcanic landscapes.

Lodgepole pine (*Pinus contorta*) in pure or nearly pure stands is widely distributed in the Pacific Northwest. This tree rapidly invades burned or logged areas and often succeeds ponderosa pine. East of the southern Cascades, however, lodgepole pine is the edaphic or topoedaphic climax on vast areas of aerially deposited pumice. L. A. Volland estimated that there were about 2 million acres of this type.[24] These soils can either have a high water table or be located in depressions where cold air is trapped, or both. C. T. Youngberg and W. G. Dahms identified 17 lodgepole pine communities in the area and considered 8 of them to represent climax situations.[25] Three of the communities include antelope bitterbrush:

Pinus contorta/Purshia tridentata
Pinus contorta/Purshia tridentata–Ribes cereum
Pinus contorta/Purshia tridentata/Festuca idahoensis

Paul Edgerton and his colleagues conducted a study on the effect on bitterbrush of logging and slash disposal of lodgepole pine in a *Pinus contorta/Purshia tridentata/Stipa occidentalis* community south of Bend, Oregon.[26] Apparently, Youngberg and Dahms did not consider this to be a climax community. David Perry and James Latan reported a lodgepole pine/antelope bitterbrush community in the west Yellowstone area on loamy sands derived from recent volcanic activity, situations similar to those previously described for Oregon.[27]

To the east of the lodgepole pine region of south-central Oregon are a series of antelope bitterbrush communities in the far northwestern Great Basin within the range of the Silver Lake mule deer herd. This is an area of transition from pine woodlands to the shrub steppe of the Great Basin. J. E. Dealy described *Purshia tridentata/Festuca idahoensis* associations here similar to those previously described farther north.[28] This community follows the margin of the pine woodlands and is transitional to the shrub-steppe communities. Dealy also described *Purshia tridentata–Artemisia arbuscula/Stipa thurberianum* communities in which the antelope bitterbrush and low sagebrush occupy different soils, in a mosaic. The low sagebrush occupies portions of the site with a layer restrictive to rooting 40–50 cm below the surface, and the antelope bitterbrush occurs on portions of the site without a restrictive layer. This community had been previously reported as occurring in the southern Blue Mountains and the southeastern Wallowa Mountains, on winter ranges for the Silver Lake mule deer herd.[29]

At higher elevations where the mule deer range in the summer months, Dealy found a *Pinus ponderosa/Purshia tridentata/Festuca idahoensis* ecosystem similar to those previously mentioned for the more northern trans-Cascade Mountains (Fig. 3.4). Antelope bitterbrush had 100 percent constancy in the examples of this classification unit sampled near Silver Lake. There were five other shrub species present in these communities as well, but none of them occurred with anywhere near the same constancy, and none was preferred by mule deer. Dealy also described a *Pinus ponderosa/Purshia tridentata/Festuca idahoensis–Carex rossii* variant of this community. Besides the obvious addition of *Carex rossii* to the understory, the number of ponderosa pine stems per acre is greatly reduced in this community, and the dominance of antelope bitterbrush is more pronounced. Dealy believed that this variant might be a wildfire-induced disclimax. At higher elevations Dealy found a *Pinus ponderosa/Arctostaphylos patula/Festuca idahoensis* community with 100 percent

Fig. 3.4. Jeffrey pine–ponderosa pine woodland with mountain big sagebrush–antelope bitterbrush understory in the volcanic highlands north of Frenchman's Lake in Plumas County, California.

constancy of antelope bitterbrush. In drier portions of the higher woodlands, *Cercocarpus ledifolius* replaces antelope bitterbrush. At even higher elevations, the ponderosa pine woodlands grade into *Pinus ponderosa–Abies concolor/Festuca idahoensis* communities where antelope bitterbrush has more than 80 percent constancy. Dealy also recognized a *Pinus contorta/Festuca idahoensis* community with more than 80 percent constancy of antelope bitterbrush similar to one described by Volland.

The Silver Lake region is not the first shrub-steppe region with antelope bitterbrush you will encounter as you proceed from north to south through the Pacific Northwest. The Columbia Basin of eastern Washington and adjacent Oregon includes some 6 million acres of steppe and shrub-steppe vegetation. This area was extensively studied by Daubenmire, and this is the environment where he developed much of his ecological classification system. He described nine zonal associations. Antelope bitterbrush here has a constancy of less than 10 percent in *Artemisia tridentata/Agropyron spicatum* associations and 80 percent in *Purshia tridentata/Festuca idahoensis* communities. The antelope bitterbrush Idaho fescue communities occur on sites with more effective moisture (i.e., moisture useable by plants) than the modal big sagebrush/bluebunch wheatgrass communities.

Within the Columbia Basin, on areas of coarser-textured soils, needle-and-thread-grass (*Stipa comata*) replaces bluebunch wheatgrass in both big sagebrush and antelope bitterbrush communities. These are the famous antelope bitterbrush stands growing on sand along the Columbia River near Boardman, Oregon. The sand-textured soils have significantly less moisture-holding capacity than the modal big sagebrush or antelope bitterbrush communities. The understory dominant grass changes, but the shrub layers remain remarkably similar across the changing soils. W. H. Rickard and R. H. Sauer described primary production in this type of community at Richland, Washington, where the understory had been converted to *Bromus tectorum* dominance.[30] Dixie Smith described a big sagebrush/bitterbrush/needle-and-thread-grass community in the Jackson Hole area of Wyoming.[31]

In eastern Oregon and extending into northeastern California and southwestern Idaho there is a well-developed western juniper zone that Franklin and Dyrness considered to be a northwestern extension of the pinyon-juniper woodlands of the Great Basin and Colorado Plateau (Fig. 3.5). As a type of vegetation, the western juniper zone may reflect a relationship with the pinyon-juniper zone, but the two may share little in their recent evolutionary history. Driscoll described five western juniper associations for central Oregon.[32] Antelope bitterbrush had 100 percent constancy in three of them:

Juniperus occidentalis/Festuca idahoensis
Juniperus occidentalis/Artemisia tridentata/Agropyron spicatum
Juniperus occidentalis/Artemisia tridentata–Purshia tridentata

In a *Juniperus occidentalis/Artemisia tridentata/Agropyron spicatum–Astragalus curvicarpus* association, antelope bitterbrush occurred in about a third of the stands sampled. In a study of a relict area in central Oregon that supports three different western juniper communities, Driscoll suggested that the distribution of antelope bitterbrush corresponds to the occurrence of coarse-textured stony soils overlying cracked bedrock.[33]

The southeastern portion of Oregon and a strip around the western and eastern sides of the Blue and Wallowa Mountains are characterized by shrub-steppe vegetation. These environments are similar but not identical with the shrub steppe of the Columbia Basin. The high deserts of southeastern Oregon are, as the name implies, much higher in elevation. Soils here tend to be shallower and less well developed, and the shrub steppe of the Columbia Basin is surrounded by a sod-forming grass meadow steppe that is generally absent in the high desert region. Much of southeastern Oregon is a northwestern extension of the

Fig. 3.5. Cattle grazing in western juniper/antelope bitterbrush–mountain big sagebrush/bluebunch wheatgrass community above Doyle Crossing, Lassen County, California.

Great Basin. N. E. West distinguished between sagebrush steppe and Great Basin sagebrush environments.[34] The Great Basin sagebrush communities occur in the central Great Basin and extend to the Colorado Plateau. The perennial grass portion of these communities is more poorly expressed and more subject to shrub dominance than the shrub steppe found to the north. The *Artemisia* communities in southeastern Oregon are transitional from shrub steppe to Great Basin sagebrush.

In the high deserts of Oregon the shrub-steppe communities have four dominant sagebrush species: *Artemisia tridentata, A. arbuscula, A. rigida,* and *A. cana.*[35] Dealy described *Artemisia tridentata–Purshia tridentata/Festuca idahoensis* communities in the Silver Lake area, as mentioned above. Dick Eckert and Paul Tueller also mentioned similar communities near Burns, Oregon.[36]

Ronald Tew described antelope bitterbrush distribution and habitat classification in the Boise National Forest of Idaho, where antelope bitterbrush occurs abundantly at elevations between 3,200 and 4,500 feet.[37] In a study conducted at Reynolds Creek in the Owyhee Mountains of southwestern Idaho, J. A. Young et al. found that antelope bitterbrush is not a high-elevation species at that latitude.[38] Far to the south, much of the Jeffrey pine–desert bitterbrush woodlands of the volcanic highlands south of Mono Lake are at elevations between 7,000 and 8,000-plus feet. Tew indicated that hot, south-facing slopes have the purest stands of antelope bitterbrush. On sites having deep, productive soils, the antelope bitterbrush plants tend to be treelike, and on harsher sites they are prostrate. He described seven communities containing antelope bitterbrush:

Artemisia arbuscula
Artemisia tridentata subsp. *vaseyana/Agropyron spicatum*
Artemisia tridentata subsp. *vaseyana/Festuca idahoensis*
Artemisia tridentata subsp. *tridentata*
Artemisia tridentata subsp. *xericensis*
Artemisia tridentata subsp. *spiciformis/Bromus carinatus*
Artemisia sp.–*Purshia tridentata/Aristida longiseta*

Tew described the low sagebrush sites as having shallow, rocky, clay-loam soils that are saturated for brief periods in the spring. The antelope bitterbrush plants are sparsely distributed and mainly decadent. This description would fit similar communities at Silver Lake, Oregon, and Bass Ridge in Lassen County, California. The Boise National Forest antelope bitterbrush communities occur on soils derived from granitic or volcanic rocks.

Rock Creek is in the north-central Great Basin in the far northeastern portion of Elko County, Nevada. Wilbert Blackburn et al. enumerated the plant communities of this watershed.[39] Those containing antelope bitterbrush include:

Amelanchier pallida/Artemisia tridentata/Agropyron spicatum
Artemisia nova/Poa secunda
Juniperus osteosperma
Juniperus osteosperma/Artemisia tridentata/Stipa comata
Symphoricarpus longiflorus–Artemisia tridentata–Amelanchier pallida/ Festuca idahoensis
Artemisia tridentata/Agropyron smithii

The big sagebrush/western wheatgrass (*Agropyron smithii* [*Pascopyrum smithii*]) community that contains antelope bitterbrush is interesting because it is one of the few examples in the Great Basin of antelope bitterbrush occurring in a community with a rhizomatous grass. The serviceberry (*Amelanchier pallida*) and snowberry (*Symphoricarpus longiflorus*) communities are interesting because they often occur in mountain brush communities in the Great Basin. These communities occur *above* the pinyon-juniper zone in an environment that contains many herbaceous and woody species that suggest a three-needle pine or mixed conifer woodland. Dwight Billings noted the absence of such woodlands in the western Great Basin many years ago. The mountain brush environments are a third landscape for antelope bitterbrush, in addition to lower-elevation sagebrush steppe and Great Basin sagebrush and pinyon-juniper woodlands. Paul Tueller and Dick Eckert studied 60 stands of the mountain brush type and collected detailed data on 37 stands. *Purshia tridentata* had 40 percent constancy in *Artemisia tridentata* subsp. *vaseyana–Symphoricarpos oreophilus/Festuca idahoensis* associations.[40] In the Pacific Northwest, northeastern California, and the Rocky Mountain states, winter range is often the limiting habitat for mule deer populations. In the more arid Great Basin, summer ranges, usually reflected by mountain brush habitats, are often the limiting or a limiting environment.

In the volcanic highlands of northwestern Nevada, Robert Berg determined that the most important antelope bitterbrush community for the Fox Mountain mule deer herd was *Artemisia tridentata–Purshia tridentata/Festuca idahoensis*.[41] This is, of course, a familiar community in the Pacific Northwest.

In all of the sagebrush–antelope bitterbrush communities, from the margins of the Columbia Basin to the Great Basin, the addition of the dark, rigid green leaves of bitterbrush to the silver gray of sagebrush is the ultimate in visual contrast. Except for the flower stalks of autumn, the sagebrush portions of the communities remain seasonally constant while antelope bitterbrush leaves change with the season. The brown antelope bitterbrush leaves of late fall may not be individually pleasing, but they provide welcome contrast to the boring gray sagebrush. In the spring the displays of massed antelope bitterbrush flowers more than compensate for the autumn drabness. Sprays of golden cream antelope bitterbrush flowers silhouetted against distant snow-covered mountains are one of the glories of spring.

As should be clear from the preceding descriptions, antelope bitterbrush, desert bitterbrush, and cliffrose are often found in or adjacent to pinyon-juniper woodlands in the Great Basin. Such woodlands occupy some 46 million acres in the southwestern states. Pinyon-juniper woodlands include relatively few tree species, but stands exhibit considerable diversity in appearance and composition. Some stands have closed canopies, while others have widely spaced trees interspersed with abundant shrubs and even herbaceous vegetation. The blend of tree species ranges from all pinyon to all juniper. Stands may contain all age classes or be nearly evenly aged.

Pinyon species usually encountered in different geographic regions of these woodlands include *Pinus edulis*, pinyon; *Pinus monophylla*, single-leaf pinyon; and *Pinus cembroides*, Mexican pinyon. The juniper species include *Juniperus monosperma*, one-seed juniper; *Juniperus osteosperma*, Utah juniper; *Juniperus deppeana*, alligator juniper; *Juniperus scopulorum*, Rocky Mountain juniper; and *Juniperus erythrocarpa*, red berry juniper. Associated shrub species vary from big sagebrush and antelope bitterbrush in the Great Basin to Gamble oak (*Quercus gambelii*) in Utah to point-leaf manzanita (*Arctostaphylos pungens*), waveyleaf oak (*Quercus grisea*), skunk-bush (*Rhus trilobata*), and blue yucca (*Yuca baccata*) in the Southwest.[42]

As a student of Paul Tueller, Robert H. Berg evaluated eight mule deer ranges in the Great Basin during the 1960s.[43] At Paine Springs in White Pine County, east-central Nevada, Berg found several communities in the pinyon-juniper woodlands. Most had black sagebrush (*Artemisia nova*) shrub layers, and some contained antelope bitterbrush. Paine Springs is an area where mule deer populations reached extreme highs during the 1940s.[44] In many of these communities the shrub and herbaceous understories were largely eliminated by overbrowsing. The health of the antelope bitterbrush in the principal communities Berg looked at was terrible: 49 percent of the shrubs were decadent, and nearly 50 percent were dead.

In interpreting successional trends in pinyon-juniper woodlands in the Great Basin, it is important to realize that many of these woodlands were extensively logged for charcoal production during the nineteenth-century silver-mining boom.[45] We discuss such impacts on browse production in a later chapter.

Farther north in the Great Basin, at the Fort Ruby exclosure, Berg found similar black sagebrush–pinyon-juniper woodland and woodland margin communities that contained antelope bitterbrush. At this location he also found a big sagebrush–antelope bitterbrush/needle-and-thread-grass community. To the south in Lincoln County, Nevada, Berg described plant communities at Blythe Springs that contained antelope bitterbrush, desert bitterbrush, and cliffrose in pinyon-juniper woodlands. Portions of this study area had been mechanically treated with an anchor chain to reduce the pinyon-juniper overstory. In the unchained woodlands, many of the *Purshia* plants were dead or dying. Obviously, black sagebrush communities in and at the margins of pinyon-juniper woodlands are important in the distribution of antelope bitterbrush in the Great Basin.

The Morey Bench in south-central Nevada is one of the first locations where cliffrose–desert bitterbrush hybrids were described (see Chapter 1). Tueller et al. mentioned, but did not describe, an *Artemisia tridentata–Purshia glandulosa–Prunus andersonii* community in this area.[46] Berg described several communities at the Morey Bench Peak location that supported desert bitterbrush. Usually, the desert bitterbrush occurred in association with needle-and-thread-grass and *Rhus trilobata*. These communities were often near, being invaded by, or in pinyon-juniper woodlands. Cliffrose also occasionally occurred in these communities.

In the eastern Great Basin, antelope bitterbrush and cliffrose extend their distribution into the mountains of Utah and onto the Colorado Plateau.[47] Desert bitterbrush is restricted to the far southwestern portion of Utah.

Arid highlands in the southwestern United States and adjacent Mexico usually feature a ponderosa pine woodland above the pinyon-juniper zone.[48] The shrub understory in pinyon-juniper woodlands generally decreases in southeastern Arizona and in the Sierra Madre Occidental in northern Mexico. In 1987 W. H. Moir and J. O. Carleton estimated that some 70 plant associations and more than 280 community types occurred in the pinyon-juniper woodlands of the Southwest, but there was no correlation of the classification systems between the Southwest and the Pacific Northwest.[49] Among the woodland associations they listed were *Pinus edulis/Purshia tridentata* and *Pinus edulis/Cowania mexicana.*

In comparison with the woodland and steppe antelope bitterbrush communities of the Pacific Northwest, very little has been written about the desert bitterbrush communities of the Southwest. Eamor Nord, who offered only a brief paragraph on California desert bitterbrush communities in his 1959 research note, considered the transition from antelope to desert bitterbrush to occur north of Mono Lake on the trans–Sierra Nevada. Nord further reported that desert bitterbrush

> is found over 1 million acres, most infrequently on alluvial fans adjacent to the Sonoran desert, at elevations between 3,500 and 10,000 feet. Annual rainfall averages between 4 to 12 inches. Desert bitterbrush is commonly associated with sagebrush, black brush (*Coleogyne ramosissima*), single leaf pinyon, and California juniper (*Juniperus californica*). Sometimes it occurs with Joshua tree (*Yucca brevifolia*), and creosote bush (*Larrea divaricata*), and honey mesquite (*Prosopis divaricata*), and mesquite (*Prosopis chilensis*). Communities of desert bitterbrush are infrequent and often isolated.[50]

Jack Brotherson has been involved in most of the community ecology research that has been published on cliffrose.[51] In central Utah cliffrose is characteristically found associated with limestone areas with west to southwest exposure at elevations between 3,900 and 7,800 feet. Cliffrose is not restricted to limestone soils. In fact, it occurs on a wide variety of soils, but usually in drought-prone exposures. Most of the soils where cliffrose grows are low in macro- and micronutrients. The most frequent herbaceous plant encountered in Utah cliffrose communities is the exotic annual cheatgrass.

Two things should be apparent from the descriptions of *Purshia* communities in this chapter. First, existing communities containing antelope or desert bitterbrush or cliffrose are often obviously in the process of changing. Second, the stands of bitterbrush that supported the vast expansion of mule deer herds during the first half of the twentieth century may have been the product of human disruptions of the environment.

Chapter Four

Ecophysiology of *Purshia*

The term *ecophysiology* is scientific jargon, but useful if properly defined. We interpret it to mean plant physiology as it applies to field situations in terms of functions that allow the plant to survive, grow, reproduce, and renew stands. Much has been written about the physiology of *Purshia*, especially *Purshia tridentata*. The classic paper on the subject is Eamor C. Nord's "Autecology of Bitterbrush in California,"[1] but there have been several lengthy Ph.D. dissertations on aspects of the physiology of antelope bitterbrush as well as numerous journal articles. Nord cited August L. Hormay as the pioneer in physiological research on *Purshia*, and certainly Hormay's "Bitterbrush in California" (1943) contains many of the first published observations on antelope bitterbrush.[2] Both Hormay and Nord treated the autecology of antelope bitterbrush comprehensively as it was known at the time their work was published. Because we treat seed physiology, granivore interactions, fire, and nutrition in separate chapters, we will encounter these two pioneer researchers again.

Hormay's phenology table for antelope bitterbrush growing in ponderosa pine woodlands in northeastern California is reproduced here as Table 4.1. Hormay found that flowering occurred on second-year wood, that twig growth began *after* flowering, and that twig growth continued into September. He believed that the late-season twig growth observed at this location was one reason why browse of this species was preferred in the fall, and observed that overutilization of twig growth inhibited flowering the next year.

Nord compared the phenology of *Purshia* plants from Inyo, Lassen, and Modoc Counties in California (Table 4.2) and found variation between stands located at different latitudes in early rather than late growth stages. Plants at northern latitudes completed their seasonal development within a shorter period. Regression analysis between the number of days from flowering to seed maturity and degrees of latitude showed that the rate of plant development changed about three days for each degree of latitude. Using data from California and Idaho, Nord showed that this relationship existed for about 10 degrees of latitude.[3]

Nancy Shaw and Steve Monsen studied the phenology of antelope bitterbrush, desert bitterbrush, cliffrose, and Apache plume accessions planted in a common garden near Boise, Idaho (Table 4.3).[4] As part of their study they

TABLE 4.1.
Seasonal Growth and Development of Bitterbrush in Lassen National Forest, 1942

Stage	May	June	Aug.	Sept.	Oct.	Nov.	Development period
Leaves growing	——						May 8–20
Plants flowering			——				June 1–20
Twigs growing		——	——	——	——		June 10–Sept. 6
Seeds ripe and falling			——				Aug. 1–10
Leaves falling				——	——	——	Sept.–Nov.

Adapted from Hormay, "Bitterbrush in California" (Res. Note 34, USDA, Forest Serv., Berkeley, Calif., 1943).

TABLE 4.2.
Phenology of Purshia *from Southeastern to Northeastern California*

Site	Wells Meadow	Doyle	Flukey Springs
County	Inyo	Lassen	Modoc
Latitude	37°26′	40°00′	41°37′
Elevation (feet)	4,450	4,250	4,400
Annual precipitation (inches)	6	10	12
Year of record	1954	1954	1954
Growth stage			
Leaves showing	March 3–12	March 15–20	April 15–May 10
Flowering	April 4–30	April 30–May 6	May 12–24
Twig growth initiated	May 5	May 5	May 20
Seed mature	July 1–10	June 30–July 8	July 10–24
Mean no. days flowering to mature seed	73	62	58

Adapted from Nord, "Autecology of Bitterbrush in California," *Ecol. Monogr.* 35 (1965): 307–334.

TABLE 4.3.
Cliffrose, Apache Plume, and Bitterbrush Accessions Grown in a Common Garden in Boise, Idaho

Species-Accession	Collection (site)	Elevation (feet)	Annual Precipitation (inches)
Cowania mexicana	American Fork, UT	5,100	15.0
Fallugia paradoxa	Richfield, UT	5,300	8.3
Purshia glandulosa	Benton Hot Springs, CA	5,700	5.0
	Snow Canyon, UT	3,800	7.1
Purshia tridentata	Janesville, CA	4,380	16.9
	Maybelle, CO	6,300	11.0
	Lucky Peak Res., ID	3,200	16.9
	Starvation Canyon, UT	7,000	23.2
	Eureka, UT	6,400	15.4

Adapted from Nancy L. Shaw and Stephen B. Monsen, "Phenology and Growth Habits of Nine Antelope Bitterbrush, Desert Bitterbrush, Stansbury Cliffrose, and Apache Plume Accessions," in *Proceedings of the Symposium on Research and Management of Bitterbrush and Cliffrose in Western North America,* ed. A. T. Tiedemann and K. L. Johnson, 55–69 (Gen. Tech. Rep. 152, USDA, Forest Serv., Odgen, Utah, 1983). 1982.

collected phenological data from two replications of 25 plants each for two years (Table 4.4). Initial leaf growth began between March 17 and 23 for all antelope bitterbrush accessions in both years. Their discovery that shoot leader growth of all accessions was initiated during the first week of May at approximately the time flower buds began to open apparently confirms Hormay's observations. Leader growth was later in 1980 than in 1979. Shaw and Monsen did not present data regarding when shoot growth terminated, but they did mention that growth slowed in July following seed dispersal and was minimal during late summer and fall. Under apparent moisture stress in late summer, all the antelope bitterbrush accessions dropped their leaves. We surveyed antelope bitterbrush during the rainless summer and fall (more than 130 days) of 1995 in the western Great Basin and observed that virtually all of the leaves were brown or had fallen by October.

The cliffrose, desert bitterbrush, and Apache plume accessions surveyed by Shaw and Monsen began leaf development earlier than the antelope bitterbrush. The leaves of cliffrose and desert bitterbrush persisted into the fall without the moisture-stress-associated leaf fall of antelope bitterbrush. The persistence or loss of leaves is an important aspect of browse production.

The leaves of *Purshia* species are highly variable in upper and lower surface color, pubescence, occurrence of glands, and shape. Generally, the leaves are much more digestible than the twigs. Therefore, leaves that persist in the fall and early winter constitute a valuable, digestible source of browse. Hormay described the leaves as ¼–¾ inch long and three-toothed on the end. The leaves are grouped in small bunches on the twigs and stems. He also reported commonly finding plants with different-colored leaves growing side by side. The lighter leaves were densely covered with matted hairs, whereas the dark ones were nearly smooth and often much smaller. Leaves on the darker-leaved plants were sticky and had a strong, obnoxious (to humans), resinous scent because of the presence of glandular hairs and pustulelike glands. Jean Alderfer described the leaves of antelope bitterbrush on older stem spurs as occurring in clusters and found size and shape variation in leaves from different locations on the *same* plant in addition to differences among plants.[5]

The leaves of desert bitterbrush are evergreen, nearly hairless, and dotted with impressed glands on the in-rolled edges. The tip of each leaf is cleft into three to five lobes, in contrast to the consistent three lobes of antelope bitterbrush leaves.[6] Nord described the color of desert bitterbrush leaves as bright green to gray and those of antelope bitterbrush as light gray-green. Obviously, he did not consider leaf color a very important or consistent characteristic for use in identifying the plants. Nord did suggest that antelope bitterbrush leaves are always three-cleft while those of desert bitterbrush have three to five lobes, and he described the glands on antelope bitterbrush leaves as stalked and those of desert bitterbrush as depressed. In general, the leaves of desert bitterbrush have much less pubescence on either surface than antelope bitterbrush leaves have.

The evergreen leaves of cliffrose are clustered along the branchlets and are ½–1 inch long. The leaves have revolute margins and a five- to seven-toothed apex. They are light to dark green, dotted with glands above, and more or less tomentose beneath.[7]

Shaw and Monsen described the flowers of antelope bitterbrush as regular, perfect, and perigynous (i.e., the flower parts are borne around the ovary rather than beneath it).[8] The stamens and corolla are inserted on the floral tube. The flowers are approximately ⅓ inch in diameter with showy, cream-colored, deciduous petals (Fig. 4.1). The fruit is a cartilaginous (leatherlike), pubescent achene with a persistent style that is papery at maturity. The flowers of desert bitterbrush are morphologically similar to those of antelope bitterbrush but larger (ca. ¾–1 inch in diameter). Shaw and Monsen described desert bitterbrush

TABLE 4.4.
Phenology of Cliffrose, Apache Plume, and Bitterbrush Accessions Grown in a Common Garden in Boise, Idaho

Species-Accession	Year	Leaf Growth Initiated	First Leaf Expanded	Flora Bud Visible	Anthesis	Leader Started	Fruit Mature
Cliffrose							
American Fork, UT	1979	April 26	May 18	May 14	June 5	May 12	July 20
	1980	April 25	May 15	April 28	June 2	May 11	July 22
Apache plume							
Richfield, UT	1979	April 18	May 4	May 22	June 10	May 5	July 28
	1980	April 16	May 2	May 5	May 25	Apr. 28	July 30
Desert bitterbrush							
Benton Hot Spr., UT	1979	April 26	May 18	April 28	May 24	May 18	July 18
	1980	April 23	May 15	April 21	May 17	May 14	July 22
Snow Canyon, UT	1979	April 26	May 18	April 28	May 25	May 20	July 18
	1980	April 23	May 15	April 21	May 21	May 14	July 21

TABLE 4.4. *(continued)*

Species-Accession	Year	Leaf Growth Initiated	First Leaf Expanded	Flora Bud Visible	Anthesis	Leader Started	Fruit Mature
Antelope bitterbrush							
Janesville, CA	1979	March 21	April 18	April 21	May 14	May 5	July 7
	1980	March 17	April 20	April 7	May 3	May 8	July 10
Maybelle, CO	1979	March 21	April 22	April 21	May 24	May 5	July 7
	1980	March 17	April 23	April 14	May 4	May 8	July 10
Lucky Peak, ID	1979	March 21	April 18	April 24	May 13	May 5	July 7
	1980	March 17	April 15	April 22	May 6	May 8	July 10
Starvation Canyon, UT	1979	March 21	April 18	April 21	May 14	May 5	July 7
	1980	March 20	April 20	May 4	May 4	May 5	July 10
Eureka, UT	1979	March 21	April 22	April 21	May 16	May 5	July 7
	1980	March 23	April 23	April 14	May 8	May 8	July 10

Adapted from Shaw and Monsen, "Phenology and Growth Habits."

Fig. 4.1. Antelope bitterbrush plant in full flower.

flowers as normally having two to three pistils, but the two accessions they planted in their common garden were markedly different, with the Benton Hot Springs and Snow Canyon accessions having a much greater number of pistils for the two years of the study. Desert bitterbrush achenes are slightly smaller and have a more pronounced beak than those of antelope bitterbrush; the developing seeds are more yellow-green, and mature seeds more reddish purple. In taxonomic treatments that consider cliffrose and desert and antelope bitterbrush to be species of *Purshia,* the morphological characteristic used to separate them is the number of pistils (*Purshia mexicana,* four to five; *Purshia tridentata,* one or sometimes three).[9] Shaw and Monsen found that 100 percent of the flowers of cliffrose and Apache plume accessions in their common garden had multiple pistils. Desert and antelope bitterbrush, on the other hand, varied considerably, with 1–28 percent of the flowers having multiple pistils. The percentage varied from one year to the next, but the accessions with high or low numbers of multiple pistils remained similar.

Shaw and Monsen found big differences between years in the dates of anthesis and fruit maturity in Idaho (see Table 4.4); however, there was virtually no difference among accessions for either year. The time between first flower bud appearance and fruit maturity was 74–94 days for the antelope bitterbrush accessions (Table 4.5). The local Idaho accession required the least time in both years for fruit maturity. The authors did not find a lot of variability in floral phenology. Nord developed a very accurate regression to predict fruit maturity by adding or subtracting one day for each 100 feet in elevation, 15 minutes of latitude, and 1¼ degrees of longitude. When longitude was dropped from the equation, the other two factors accounted for 74 percent of the variation.[10]

Shaw and Monsen used marked flowers to follow reproductive success. On June 30, 1980, 38–55 percent of the marked flowers of antelope bitterbrush were producing normal achenes. By July 10, fully developed achenes were collected from only 14–24 percent of the marked flowers. Wind and rain can cause severe losses of maturing achenes. Once the fruits are fully mature they dehisce very quickly.

Frank Stanton's Ph.D. dissertation includes a phenology table for antelope bitterbrush (Table 4.6).[11] His format is very similar to that used by Hormay, although Stanton added three elevation classes. For his low-elevation populations he used stands along the Columbia River between Dallas and McNary. Data for the intermediate elevations came from the lower eastern slopes of the Cascade Range in the Metolius-Sisters and Warm Springs areas. The high-elevation areas were in lodgepole and ponderosa pine woodlands in the Paulina Mountains south of Bend, Oregon. Stanton observed that all antelope bitterbrush plants in a given stand appeared to flower simultaneously and the seeds to mature simultaneously. He also observed, as did Shaw and Monsen, that nearly the entire crop of achenes could be shattered to the ground in a single afternoon by a windstorm. Stanton suggested that as a rule of thumb, antelope bitterbrush seeds mature seven weeks after flowering, but his meaning is not entirely clear. If he meant seven weeks from the date when flowers dry until the fruit matures, his 49 days would fall within the 42–60-day range for fruit development that Shaw and Monsen observed in their Boise common garden (see Table 4.5). Stanton observed the phenology of low-elevation stands along the Columbia River for four years. During that time the flowering date occurred within the same two-week period every year (April 15–29). Seed-harvesting dates in these stands were June 15–23, 1951; June 16–20, 1952; June 17, 1954; July 5, 1955; and June 9, 1958. J. P. Blaisdell reported that seeds of the antelope bitterbrush in the eastern Snake River Plains all matured within a six-day period.[12]

The quantity of seeds produced varies greatly among years, species, and locations. Nord considered desert bitterbrush seed production in California to be much more variable than that of antelope bitterbrush because of the more arid

TABLE 4.5.
Duration of Flowering and Fruiting Phases for Cliffrose, Apache Plume, and Bitterbrush Accessions

Species-Accession	Year	Flowering Period (days)	Fruit Development (days)	Total (days)
Cliffrose				
American Fork, UT	1979	30	37	67
	1980	47	38	85
Apache plume				
Richfield, UT	1979	31	36	67
	1980	35	57	86
Desert bitterbrush				
Benton Hot Spr., UT	1979	31	50	81
	1980	45	47	92
Snow Canyon, UT	1979	30	51	81
	1980	44	47	91
Antelope bitterbrush				
Janesville, CA	1979	33	44	77
	1980	37	57	94
Maybelle, CO	1979	33	33	77
	1980	28	59	87
Lucky Peak, ID	1979	30	44	74
	1980	24	55	79
Starvation Canyon, UT	1979	34	43	77
	1980	27	60	87
Eureka, UT	1979	35	42	77
	1980	28	59	87

Adapted from Shaw and Monsen, "Phenology and Growth Habits."

locations occupied by desert bitterbrush. Obviously, if *Purshia* plants flower on second-year wood, the leader growth the previous year greatly influences seed production. Current seasonal growth and fruit development occur right at the time when bitterbrush is most dependent on stored carbohydrates. Nord suggested these relations before portable equipment was available for measuring photosynthetic activity in the field, basing his thesis on generalized data for range forage species developed by E. C. McCarty and R. Price.[13] When B. R. Mc-

TABLE 4.6.
Phenology of Antelope Bitterbrush in Oregon

Phenological Stage	Elevation (feet)	Feb.	Mar.	Apr.	May	June	July
Leaf	200–600	X	XXXX	X	—	—	—
development	2,500–3,500	—	X	X	—	—	—
	4,000–5,500	—	—	XX	—	—	—
Flowers mature	200–600	XX	X	—	—		
	2,500–3,500	—	—	—	XXXX	—	—
	4,000–5,500	—	—	—	X	XXX	—
Seeds mature	200–600	—	—	—	—	XXX	X
	2,500–3,500	—	—	—	—	—	XXX
	4,000–5,500	—	—	—	—	—	XX

Adapted from Frank W. Stanton, "Autecological Studies of Bitterbrush (*Purshia tridentata* [Pursh] DC.)" (Ph.D. diss., Ore. State Univ., Corvallis, 1959).

Connell and G. A. Garrison analyzed seasonal variations of available carbohydrates in antelope bitterbrush in Colorado in the 1990s, they found that Nord was right. Both the percentage and the weight of total available carbohydrate compounds were cyclical for all portions of the plant.[14] Carbohydrate accumulations decreased during the rapid twig growth and seed maturation period and remained low until mid-August. In the early fall, reserves increased until leaf fall occurred. Antelope bitterbrush subjected to two moderate defoliations during quiescence and rapid growth (or flower development) retained fair to good vigor at the end of the growing season.[15] Basal stem total nonstructural carbohydrates were severely reduced below those of control plants by all defoliation treatments.

Nord noted that increased photosynthetic activity resulted in increased root storage of carbohydrates and greater potential seed production the next year. He reported records of precipitation and leader growth of antelope bitterbrush with comments on subsequent seed production for his Lassen and Inyo County sites for a seven-year period (Table 4.7). Good seed crops did not coincide with, but usually followed, years of favorable precipitation. He even suggested that good seed crops could develop in years with below-normal precipitation if at least 3 inches of leader growth had occurred the previous year.

Nord estimated that antelope bitterbrush seed production could reach 500 pounds per acre in excellent stands in good years, but this estimate was based on a very small sample: 2 plants, 4 and 9 feet tall, in stands of 500 plants per

TABLE 4.7.
Antelope Bitterbrush Seed Production in Relation to the Previous Year's Precipitation and Leader Growth

Season Oct. 1– Sept. 30	Inyo County			Lassen County		
	Precip. (inches)	Leader Growth (inches)	Seed Production	Precip. (inches)	Leader Growth (inches)	Seed Production
1952–53	3.6	2.1	—	13.7	3.6	—
1953–54	6.0	2.7	poor	10.1	2.6	excellent
1954–55	5.5	1.8	fair	8.3	2.6	poor
1955–56	8.2	4.4	poor	21.1	5.8	poor
1956–57	3.7	1.7	good	14.7	5.3	good
1957–58	8.0	4.2	poor	17.9	4.2	fair
1958–59	4.2	1.8	excellent	9.7	2.5	good
averages	5.4	2.7	—	14.7	4.1	—

Adapted from Nord, "Autecology of Bitterbrush in California."

acre. He did not give the location, but the size of the shrubs suggests Lassen County. Although he conducted extensive research in northeastern California, Nord did not mention the influence of severe spring frost on antelope bitterbrush seed production. Shaw and Monsen noted that all the *Purshia* species in their Boise common garden flowered after spring frost. However, in 1979 they recorded two nights of below-freezing temperatures (May 28, 28°F; and May 29, 31°F) with no frost damage to developing fruits. We have observed several seasons along the trans-Sierra front when severe late frost completely eliminated antelope bitterbrush seed production in Janesville and Doyle, California, and Jacks Valley, Nevada.

Hormay stated that antelope bitterbrush plants in natural stands had to be 10 years old before they produced seed, but Shaw and Monsen found that plants well spaced in their common garden flowered and produced fruits at a much younger age. Big sagebrush is a major component of many antelope bitterbrush communities. Under natural conditions, big sagebrush seedlings in stagnant, overstocked stands can remain dwarf plants for many years, but spaced, planted seedlings can produce abundant seed crops in their second year. Big sagebrush plants flower at the end of the summer drought. Thanks to a leafy inflorescence that is extremely efficient in fixing carbon per unit of available moisture, they produce seeds without negatively influencing the carbohydrate balance of the plant. This may be a major factor in the ability of woody sagebrush species to attain dominance in post-Pleistocene deserts while *Purshia* species became restricted relics.[16]

Although it was not evident in Shaw and Monsen's Idaho garden, cliffrose plants in the Southwest can flower in May or June and again in August. The second flowering is in response to summer precipitation. The best seed is produced from the first flowering, but some seed may be produced in September and October from the later flowering.[17]

The wide range of habitats occupied by antelope bitterbrush communities is matched by the variation in the appearance of the plants. The exceptionally large, almost treelike plants found along the eastern slopes of the Sierra Nevada have already been mentioned. In pine woodlands, the plants are often much shorter, and with increasing elevations they often have finer stems and spreading crowns. In woodland situations antelope bitterbrush plants are often shorter than the normal depth of the winter snowpack. This is a regional adaptation. Many woodland shrubs risk frost damage if their roots are in near frozen or frozen soil while branches with persistent leaves remain above the surface of the snowpack.

Nord divided antelope bitterbrush into three size classes: (1) tall-massive, (2) intermediate, and (3) semiprostrate, and planted experimental plots using the seeds of the different growth forms. He discovered that seed from tall-massive plants always produced tall-massive plants; seed from intermediate plants produced 68 percent erect plants, 18 percent semierect plants, and 14 percent semiprostrate plants; and seed from semiprostrate plants produced 26 percent erect plants, 30 percent semierect plants, 39 percent semiprostrate plants, and 5 percent prostrate plants. The height and crown diameter of progeny from the various growth classes tended to vary at different garden locations.[18] Alderfer detected consistent differences in the growth form of seedlings grown in a common environment in the greenhouse from seeds collected from accessions with varying growth forms.[19]

Shaw and Monsen found three growth habits in the antelope bitterbrush accessions grown in their common garden as well, which they classified as erect, erect-diffuse, and diffuse. They observed that the plants had the same growth form in the garden as they exhibited in the natural communities where they were collected.[20] The erect plants typically had a small number of basal stems with new leader growth occurring from a few upper stems. The diffuse plants had a large number of basal stems, and new leaders were initiated over the entire crown.

A morphological characteristic of some higher elevation stands that has received a lot of attention is *air layering,* a form of vegetative reproduction that occurs when decumbent limbs touch the soil surface and roots develop from the stem buds. Essentially, a ring of rooted stems can develop around the original plant, although it is more common to find only one or two rooted plants next to the original. Air layering is usually associated with diffuse plants. Shaw and Monsen studied three accessions that they considered diffuse-type plants. These accessions, from Starvation Canyon and Eureka, Utah, and Maybelle, Colorado, produced plants in the common garden near Boise with 36, 0, and 8 percent air layering, respectively. The local Ada County, Idaho, source was classified as erect-diffuse and produced air layering on 2 percent of the plants grown in the common garden. No one has ever reported air layering from the erect growth form of antelope bitterbrush.

The ability to reproduce by air layering has a direct influence on the sprouting of antelope bitterbrush plants after the aerial portion of the shrub has been burned in a wildfire or prescribed burn. In the simplest interaction with wildfire, a very light burn will kill the mother plant and leave one or several of the air-layered offspring because the litter that accumulates close to the central stem of the original plant enhances fire temperatures both in degree and duration. Apparently, ecotypes that are prone to air layering have a greater potential to sprout after burning. We discuss this attribute in more detail in Chapter 10. Here it is important to know only that sprouting of antelope bitterbrush plants after wildfires is influenced by a number of factors, including the genetic attributes of the plant, the physiological status of the plant at the time of burning, the physical nature of the fire, and environmental conditions after the fire.

Supposedly, all desert bitterbrush plants sprout after burning. In addition, Nord reported rare ecotypes of desert bitterbrush at very high elevations or on very shallow soils that air-layered. Apparently, this is also a cliffrose characteristic. Nord recognized two types of sprouting by the antelope bitterbrush he studied on highway rights-of-way from Reno and Carson City, Nevada, up the eastern side of the Sierra Nevada: sprouting from stem buds and sprouting from a ring of buds on the top of the root crown. Obviously, in order for the former to occur, there must be living tissue left above the ground after the fire. Sprouting from the root crown sometimes occurred after the entire aerial portion of the plant had been consumed, but it was very rare. Nord also reported that desert bitterbrush can sprout from cut root surfaces, with soil moisture being the controlling factor.

Hormay observed significant sprouting of antelope bitterbrush only twice in California. Nord, however, studied postwildfire succession in a number of antelope and desert bitterbrush stands in California (Table 4.8). Apparently, he visited many of the wildfire sites several years after the fires occurred, and it is possible that he misidentified seedlings as sprouts. We consider the identification of sprouts versus seedlings to be a difficult task as succession proceeds

TABLE 4.8.
Sprouting of Antelope and Desert Bitterbrush Following Wildfires in California and Nevada

Name of Burn	General Location	Year Burned	Relative Frequency of Sprouting
Antelope bitterbrush			
Sheepwell	Klamath National Forest	1949	Rare
Mt. Dome	Modoc National Forest	1944	Rare
Rifle Ridge	Shasta National Forest	1954	Frequent
Lookout	Modoc National Forest	1952	Infrequent
Crater School	Modoc National Forest	1952	Infrequent
Doyle	Lassen County	1957	Rare
Susanville	Lassen County	1946	Infrequent
Crystal Peak	Plumas National Forest	1950	Frequent
Peterson Mountain	Washoe County, NV	1954	Frequent
Loyalton	Tahoe National Forest	1953	Frequent
Manke's Sawmill	Toiyabe National Forest	1950	Abundant
Desert bitterbrush			
McMurray Meadow	Inyo National Forest	1948	Abundant
George's Creek	Inyo National Forest	1949	Abundant
Walker Pass	Sequoia National Forest	1951	Frequent
Pinion	Angelus National Forest	1953	Abundant
Truck	Angelus National Forest	1947	Abundant
Doble	San Bernardino National Forest	1946	Abundant
Mica Gem	San Diego County	1954	Frequent
Mid-Hills	Providence Mountains	1955	Infrequent
Murphy Lease	Providence Mountains	1951	Frequency
Kingston	Kingston Mountains	1958	Abundant

Abundant = 25 percent or more, Frequent = 5–24 percent, Infrequent = 1–4 percent, and Rare = >1 percent.

Adapted from Nord, "Autecology of Bitterbrush in California."

after fires. Blaisdell and Mueggler reported that antelope bitterbrush on the upper Snake River Plains commonly sprouted after a wildfire or top removal.[21] An important consideration in evaluating antelope bitterbrush sprouts from burned or cut surfaces is the persistence of sprouts in subsequent seasons.[22]

Hormay, who age-dated a considerable number of antelope bitterbrush plants, was the first to suggest that growth rings offer a fairly precise estimate of age. Using that technique he estimated the common longevity of antelope bitterbrush plants in California at 60–70 years. The oldest plant he encountered had 82 growth rings on the stem at the soil surface. Nord reported a plant from Panum Crater in the volcanic highlands south of Mono Lake with 128 growth rings. Most likely this was a desert bitterbrush plant, although Nord did not refer to the Mono Lake–volcanic highland *Purshia* plants as antelope or desert bitterbrush, but instead classified them as transition plants. Nord claimed that the largest antelope bitterbrush plant on record was growing 6 miles south of Janesville, California.[23] The stem circumference was 36 inches and the plant was 12 feet tall with a maximum crown circumference of nearly 20 feet. Richard Driscoll immediately countered with a plant in Cove Palisade State Park in Jefferson County, Oregon, that was 38 inches in circumference at the stem base, 14.2 feet tall, and had a crown circumference of 21.7 feet.[24] This giant arborescent form of antelope bitterbrush also occurs at Mount Pleasant in Sanpete County, Utah.[25] Obviously, the tall, treelike growth form is not restricted to the Lassen County area.

During the 1970s and 1980s many common garden studies were initiated to study the performance of antelope bitterbrush accessions. We have already discussed Shaw and Monsen's phenological observations made on one such garden planted near Boise. A more common practice was to plant gardens and record stand establishment, growth, and persistence. J. N. Davis reported on the results of numerous plantings throughout Utah, including 141 accessions of antelope bitterbrush, 49 accessions of cliffrose, 46 accessions of antelope bitterbrush–cliffrose hybrids, and 1 desert antelope bitterbrush hybrid. Differences in stand establishment, vigor, and various growth measurements led him to conclude that only 24 of these accessions were suitable for planting in the pinyon-juniper woodland zone of Utah.[26]

P. J. Edgerton et al. established a similar garden about 30 miles east of Baker, Oregon. The Keating site was a degraded big sagebrush community that had once been an important antelope bitterbrush stand for mule deer.[27] The accessions planted in this garden are described in Table 4.9. The first season was favorable for shrub establishment, but drought set in the second growing season. The influence of the drought was compounded by a severe grasshopper infestation (seven grasshoppers per square yard). The defoliation of the antelope bit-

TABLE 4.9.
Shrub Accessions Evaluated in the Baker, Oregon, Common Garden

Species-Accession	Elevation (feet)	Precipitation (inches)	Soils	Growth Habit
Antelope bitterbrush				
Janesville, CA	4,036	10	granitic	erect
Boise Basin, ID	3,753	30	granitic	decumbent
Fort Hall, ID	3,753	17	alluvial	semierect
Garden Valley, ID	3,568	20	lacustrine	erect
Hat Rock, OR	348	7	granitic	erect
Keating, OR	2,577	6	granitic	erect
Pringle Falls, OR	4,264	20	pumiceous	decumbent
Warner Mountains, OR	5,946	20	basaltic	decumbent
Cliffrose				
American Fork, UT	4,163	15	sedimentary	erect
Apache plume				
Richfield, UT	5,252	13	sedimentary	erect

Adapted from P. J. Edgerton, J. M. Geist, and W. G. Williams, "Survival and Growth of Apache Plume, Stansbury Cliffrose, and Selected Sources of Antelope Bitterbrush in Northeastern Oregon," in Tiedemann and Johnson 1983, 44–54.

terbrush plants varied among accessions. Pringle Falls, Oregon, was particularly hard-hit by the grasshoppers. (We discuss plant-insect relations in more detail in Chapter 9.) Cliffrose and Apache plume were well north of their natural range in this common garden, but both species persisted and grew very well.

Edgerton et al. collected data on crown diameter and yearly aboveground biomass of the shrub seedlings. The average aboveground biomass for all antelope bitterbrush seedlings was 8, 70, 604, and 687 grams, for one, two, three, and four years, respectively, after the garden was established (Table 4.10). Shaw and Monsen, who also collected height and crown diameter data in their Boise common garden, considered the plants *within* accessions to exhibit great variation in growth form. If their final height and crown diameter growth figures are compared, the erect antelope bitterbrush had a height-to-crown-diameter ratio of 1:1.1. The intermediate source had a ratio of 1:1.6, and the diffuse accessions had an average ratio of 1:1.8. It is apparent from these figures that young ante-

TABLE 4.10.
Height (Inches) of Shrub Accessions in the Baker, Oregon, Common Garden

Species-Accession	1976*	1977*	1978*	1981*
Antelope bitterbrush				
Janesville, CA	10.4±0.8	18.6±0.8	32.2±1.2	35.2±1.6
Boise Basin, ID	9.6±0.8	16.4±1.2	26.0±1.2	30.4±1.2
Fort Hall, ID	8.0±2.8	12.4±2.4	20.4±2.4	24.4±1.2
Garden Valley, ID	8.4±1.2	15.2±1.2	27.6±2.0	33.2±2.4
Hat Rock, OR	9.6±2.8	16.0±1.6	27.6±2.0	33.2±2.4
Keating, OR	10.0±0.4	16.8±1.6	26.8±1.2	31.2±1.2
Pringle Falls, OR	7.6±1.2	13.6±1.2	23.2±2.0	18.8±4.8
Warner Mountains, OR	8.8±1.2	14.4±1.6	25.6±2.0	27.2±2.0
Cliffrose				
American Fork, UT	8.8±0.4	13.6±1.2	24.4±1.2	32.0±2.0
Apache plume				
Richfield, UT	18.8+1.6	23.2+1.6	32.0+2.4	41.2+2.4

* Maximum height ± 0.05 confidence interval.

Adapted from Edgerton et al., "Survival and Growth in Northeastern Oregon."

lope bitterbrush plants, even from erect forms, are globular to flat-globular in shape.

C. J. Bilbrough and J. H. Richards investigated the growth factors that determine the architecture of antelope bitterbrush plants.[28] Their "branch-level" approach makes it possible to distinguish between the plant's inherent growth qualities and environmental effects on growth. They found that shrub growth results from the repeated production of basic morphological units that are integrated in short and long shoots differentiated by the absence and presence, respectively, of internode elongation. Production of long shoots ultimately determines the size and form of the plant. Long shoots in turn are integrated into an architectural complex through mechanisms based on hormonal control. Apparently, Bilbrough and Richards studied a diffuse or decumbent form of antelope bitterbrush because the shrub architecture they described was "branched (decumbent) . . . with few long and many short shoots. The leader rarely grew and was no longer than laterals." It would be interesting to apply the same procedures to erect forms of antelope bitterbrush.

As previously mentioned, the prostrate forms of antelope bitterbrush growing on the tablelands bordering lower Willow Creek northeast of Susanville, California, are kept close-cropped by excessive browsing. The site where these plants are found is across the valley from Janesville, where the well-known erect form of antelope bitterbrush is found. The Janesville plants occasionally manage to escape extreme browsing by growing taller than browsing animals can reach. Typically, such plants exhibit a massive central trunk with the umbrella-shaped canopy above the reach of browsers.

The occasional negative height growth rates reported for young antelope bitterbrush plants are apparently caused by winter frost damage. Stanton, for example, reported that in November 1955, what the Weather Bureau termed "extremely unseasonable" cold weather damaged young plants in his nursery in eastern Oregon.[29]

Growth rates of antelope bitterbrush in garden studies apparently greatly exceed those found in natural communities where competing vegetation occurs. The plants in Shaw and Monsen's garden in Idaho grew 50 inches tall in 5 years. In natural stands in Oregon, however, Stanton found that naturally established plants in a variety of communities were only about 5 inches tall after 5 years (Fig. 4.2).[30] Stanton reported that during the 1950s Oregon Game Commission field agents regularly sampled ungrazed antelope bitterbrush growth each fall on 30 important mule deer ranges in eastern Oregon. The twig growth of approximately 4,300 plants on permanent transects was measured for 5 years, with the average annual growth being 4.1 inches. Growth rates depended on precipitation and ranged from a minimum of 2.7 inches in 1955 to 5.1 inches in 1956. Robert Kindschy measured the leader growth of antelope bitterbrush plants in southeastern Oregon over a 15-year period and determined that there was a significant ($P \geq 0.01$) relation between precipitation and leader growth, but his equation accounted for only 36 percent of the variability in browse production.[31]

Richard Hubbard and Reed Sanderson packed a great deal of information on antelope bitterbrush growth, structure, and browse production into a short research note published in 1950 that is seldom cited today.[32] They collected five maturity classes of bitterbrush from the mule deer management area at Doyle, California (Fig. 4.3). The area had been fenced to exclude livestock six years prior to the collection of the plants. Hubbard's group removed all leaders and spurs from the plants by age class. In the process a considerable number of leaves were shattered; because these could not be assigned to leaders or branch spurs, they were recorded as a separate browse class. There was no relationship between weight of leaders produced and either size or age. The plants intermediate in size and age produced the heaviest leaders. The weight of leaders produced by the largest plant was less than one-half that of the smallest plant. The largest plants had the greatest total browse production (spurs, leaders, and shat-

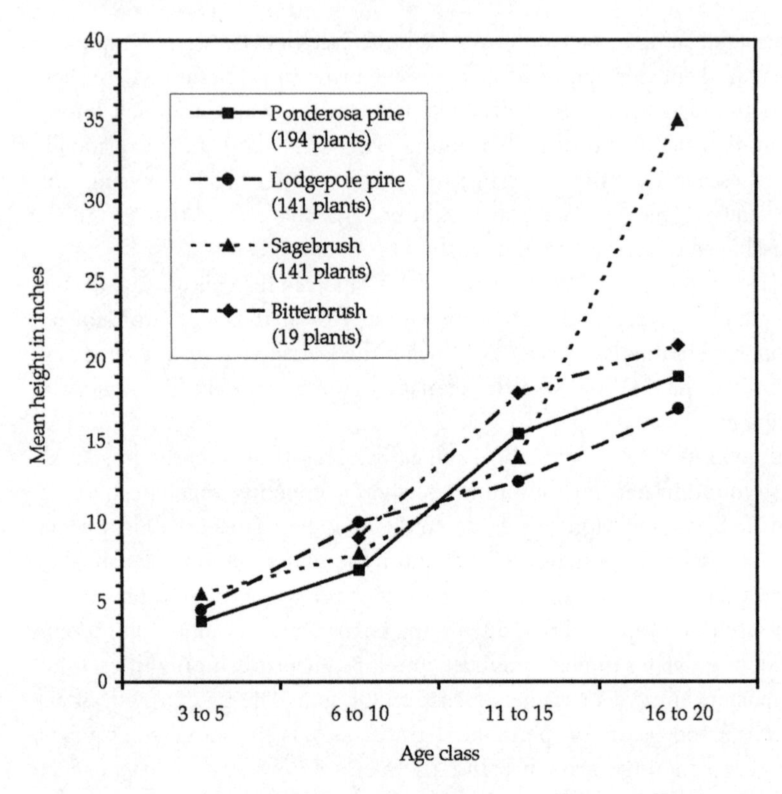

Fig. 4.2. Height of antelope bitterbrush by age classes for four different types of plant communities in Oregon. Adapted from Stanton, "Autecological Studies of Bitterbrush."

tered leaves combined); however, the total was only 1.46 pounds, compared with 1.26 pounds for the smallest plant.

The inherent mechanisms and constraints that limit or facilitate production of new growth after herbivory are not well understood. Several researchers have suggested that inherent growth rate may be the major determinant of a plant's capacity to regrow and compensate for lost photosynthetic tissue. Plants with high inherent growth rates have the potential to produce new growth rapidly and compensate for lost tissue. Wandera et al., who investigated factors influencing growth for several sagebrush-zone shrub species, found that severely clipped mountain big sagebrush (*Artemisia tridentata* subsp. *vaseyana*) plants died, while the four rose family species—antelope bitterbrush, serviceberry (*Amelanchier alnifolia*), mountain mahogany (*Cercocarpus montanus*), and curlleaf mahogany (*C. ledifolius*)—produced compensatory growth equal to the amount removed.[33] The shrub species significantly differed in meristem-

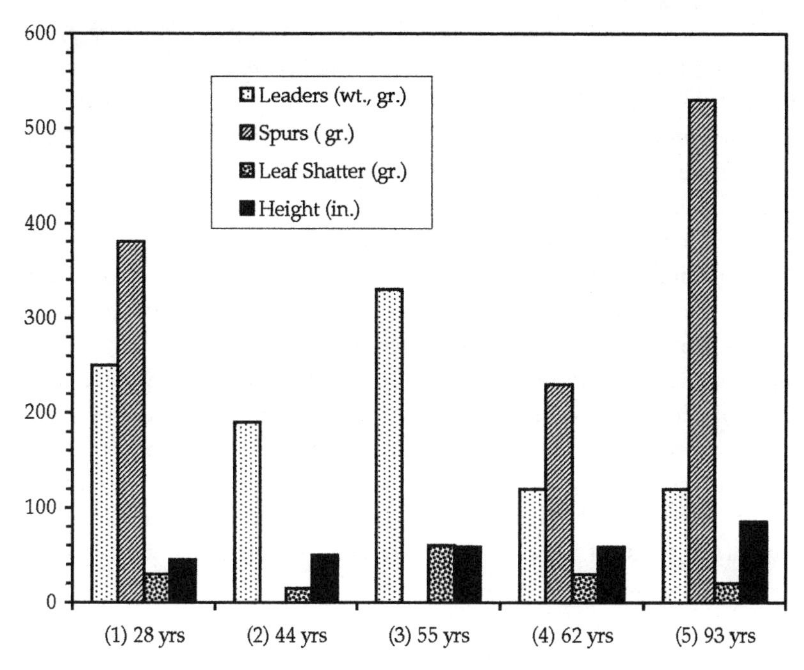

Fig. 4.3. Sizes and ages of five antelope bitterbrush plants harvested at Doyle, California, and the weight of spurs, leaders, and shattered leaves produced by each. Adapted from Richard L. Hubbard and H. Reed Sanderson, "Herbage Production and Carrying Capacity of Bitterbrush" (Res. Note 157, USDA, Forest Serv., Berkeley, Calif., 1950).

atic potential. Curlleaf mountain mahogany and serviceberry had the greatest and least number of buds and long shoots per plant, respectively. Antelope bitterbrush fell between these extremes. The number of long shoots produced following bud removal was positively correlated with the growth biomass, while the percentage of long shoots produced at the basal position on twigs was negatively related with new growth biomass. In a study published in the early 1950s, G. A. Garrison found that a single defoliation (described as removing the tips of the twigs) of antelope bitterbrush during the dormant period stimulated greater twig production compared with unbrowsed plants.[34]

An interesting side issue concerning the physiology of antelope bitterbrush is its reported use by American Indians as a medicinal plant. Percy Train, the pioneer Great Basin botanical collector, reported that Indians used decoctions (extracted by boiling) of antelope bitterbrush as remedies for venereal diseases, tuberculosis, pneumonia, and several common ailments.[35] Henry Trimble conducted preliminary experiments on the medical properties of antelope bitterbrush seeds as early as 1892 but reported nothing of significance. Using plant material collected in Washoe County, Nevada, Charles Netz conducted phytochemical analyses of antelope bitterbrush as part of his Ph.D. research at the University of Minnesota, which was sponsored by the Indian Medicinal Plant study of the Bureau of Plant Industry, USDA. He attributed the purported curative action obtained by the Indians from infusions, decoctions, and poultices of antelope bitterbrush to the bacteriostatic action of the tannins and the emulcent effect of the mucilage and pectin. Netz attributed the bitter taste of bitterbrush to the presence of a catechol or catechol-ploroglucin tannin. It would be interesting to have similar data for desert bitterbrush and cliffrose.[36]

Perhaps the most significant aspect of the ecophysiological literature concerning *Purshia* is what is missing. There are no modern field studies evaluating photosynthetic activity and water relations. Considering the importance of the *Purshia* species, this gap in the information is remarkable. Federal and state agencies that are responsible for the management of mule deer should be actively supporting modern ecophysiological research on the *Purshia* species.

Chapter Five

Purshia Seed Physiology

As soon as August Hormay began to experiment with the artificial planting of antelope bitterbrush seeds he discovered that the seeds are initially dormant.[1] He correctly described most of the characteristics of the germination of the seeds. However, some of the misconceptions about the germination ecology of this species that became fixed during those early investigations resulted in misguided research and management decisions that detracted from the goal of restoring antelope bitterbrush populations.

Hormay determined that antelope bitterbrush achenes (seeds) need to be placed in a moist environment at a temperature between about 32 and 41°F for five to eight weeks before substantial germination will occur.[2] Seeds that were not pretreated had a maximum of 20 percent germination. (In our experience the germination of untreated seeds is often around 5 percent.) Hormay referred to the necessary dormancy-breaking treatment by the German forestry term *stratification.* German foresters learned to satisfy the germination requirements of seeds with this type of dormancy by placing them in alternating layers with peat and sand in cold chambers. The resulting layers had the appearance of geologic sedimentary strata, hence the name stratification. The term is sometimes confused with scarification, a very different treatment that involves mechanically or chemically breaking the seed coat. Modern seed physiologists use the self-descriptive terms *cool-moist pretreatment* and *moist prechilling* to identify stratification treatments. Hormay emphasized the importance of soaking the seeds and placing them in a substrate that supplies constant moisture for the prechilling to be effective. Nowadays this step is often ignored, and the seeds are placed in elaborate cold-temperature storage facilities designed to break the dormancy. You can store bitterbrush seeds forever under dry conditions and not satisfy the restrictive dormancy requirements.

Hormay also determined that the embryonic antelope bitterbrush plant contained within the achene is not dormant. Rather, the dormancy is imposed by the layers of the achene that surround the true seed coat (Fig. 5.1). If the dry seeds were soaked in water to soften them, he discovered, the embryos could be removed by dissection and they would germinate. Eamor Nord later developed this "excised embryo" technique into a quick test for viability of antelope bitterbrush seeds.[3]

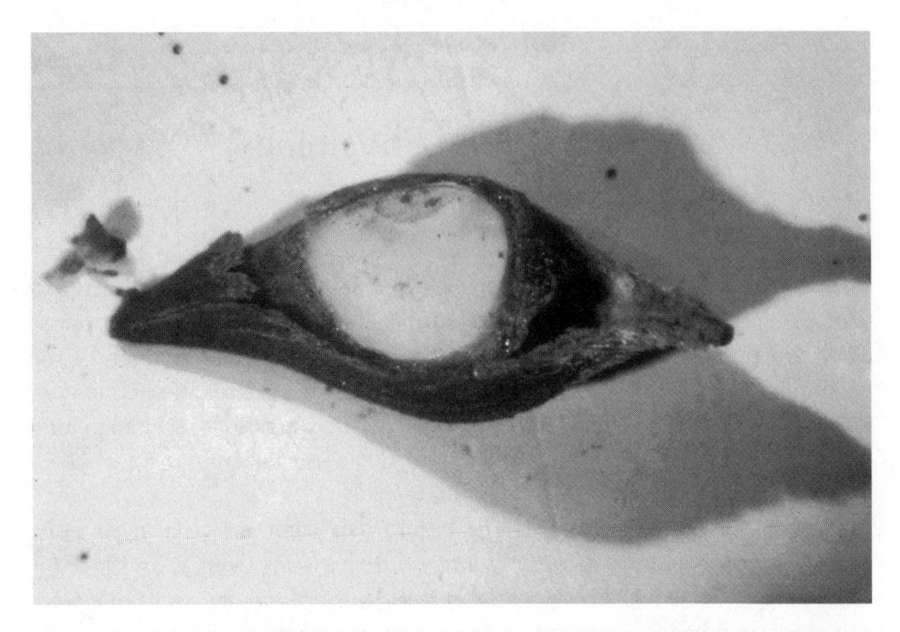

Fig. 5.1. Cross section of antelope bitterbrush achene showing the dark achene coat and white embryo. The bulk of the white interior of the achene is composed of the cotyledons, or seed leaves, which store food for the germinating seedling. The root (radicle) and hypocotyl (embryonic stem) are in the dense area nearly surrounded by the cotyledons.

Among Hormay's discoveries in the seed physiology of antelope bitterbrush fruits was the germination-inhibiting characteristics of the remnant style that covers the achenes when they dehisce from the plant. Hormay suggested that the style parts contain a powerful germination inhibitor that is not significant in the natural regeneration of antelope bitterbrush because virtually all the seeds are collected by rodents, which remove the remnant style before caching them. Hormay further reported that a large percentage of antelope bitterbrush fruits with the style or husk intact that were planted in September emerged the following spring. He interpreted this to mean that the inhibitor was either water-soluble or was destroyed by microbial decomposition during the winter. In the half-century-plus that has passed since Hormay made these observations, no one has followed this line of research to confirm the existence and elaborate on the nature of this germination inhibitor. If the inhibitor is water-soluble, it may play a role in developing seed banks of antelope bitterbrush in the absence of rodents. The style covering of cliffrose fruits is much more indurate and remains on the achene.

In the early 1970s the Wildland Seed Laboratory of the USDA, ARS, in Reno, Nevada, developed refrigerated incubation equipment so that seeds could be incubated at temperatures slightly above and below the freezing point. We used this equipment to undertake a detailed study of the dormancy characteristics of antelope bitterbrush.[4] Seeds were placed in moist blotter paper and incubated at 24, 28, 32, 36, 43, 49, and 53°F for periods ranging from 1 to 12 weeks. At 36°F it took only 3 weeks to attain 90 percent germination (Table 5.1), although pretreatment temperatures ranging from 32 to 43°F also produced satisfactory germination. If the incubation temperature was raised to 50°F or dropped to 24°F, the dormancy remained intact. From these results we can deduce that field plants require a constant environment that remains at or just above freezing for 3 or 4 weeks.

Eamor Nord suggested that a seed buried in the surface soil under continuous snow cover would receive ideal cool-moist pretreatment.[5] Rodents collect and bury antelope bitterbrush seeds in compact caches about ¾–1 inch below the soil surface. At higher elevations in ponderosa pine woodlands continuous snow cover is common. Many of the truly vital winter range areas for mule deer, however, occur at lower elevations below the continuous snow line, where conditions are not ideal for meeting antelope bitterbrush's pregermination requirements. Burial is important for germination. In a study we conducted in cooperation with the Pacific Northwest Watershed Research Laboratory in the Reynolds Creek watershed of the Owyhee Mountains of Idaho, we found that antelope bitterbrush seeds left on the surface of the soil had lower subsequent germination than seeds buried an inch below the soil surface.[6] The surface seeds were kept in nylon mesh bags for ease in recovery; bare seeds on the soil surface must be at an even greater disadvantage.

Apparently, no one has ever compared the moist-prechilling requirement of seeds from the upper portion of ponderosa pine woodlands with that of seeds from low-elevation sites. The decumbent form of antelope bitterbrush that occurs at higher elevations is covered by snow for extended periods, and it appears that once the moist-prechilling requirement is satisfied the seeds will die under laboratory conditions. Most likely, the seedlings send roots under the snow in field conditions.

Fluctuating temperatures can make the surface soil an unsatisfactory environment for antelope bitterbrush seeds, but moisture is equally important. For cool-moist pretreatment to be effective the seeds must be fully imbibed (soaked) and kept continuously in that state for the duration of the treatment. We investigated the influence of reduced osmotic and matric potentials on the cool-moist pretreatment of antelope bitterbrush seeds in the same study described above in which temperature regimes were varied.[7] Osmotic stress is a function of the potential of the solution in contact with the seed (achene) coat. A soil solution containing dissolved salts has a reduced osmotic potential, and seeds must use more energy to draw water from the solution. Antelope bitterbrush is

TABLE 5.1.

Estimated Percentage Germination of Antelope and Desert Bitterbrush and Cliffrose Seeds, Without Pretreatment[1]

16-Hours Cold-Period Temperature (°C)	8-Hours Warm-Period Temperature (°C)									
	0	2	5	10	15	20	25	30	35	40
Antelope Bitterbrush Seeds										
0	0	11±8	17±7	24±6	30±6	32±6	32±6	30±6	25±7	17±10
2	—	8±8	14±6	23±5	28±5	31±5	32±5	30±5	25±5	18±8
5	—	—	10±7	19±4	25±4	29±4	30±4	29±4	25±4	18±7
10	—	—	—	11±7	18±5	23±5	26±5	25±4	22±5	17±6
15	—	—	—	—	9±7	15±5	19±5	19±5	17±5	13±7
20	—	—	—	—	—	5±7	9±5	11±4	10±5	7±8
25	—	—	—	—	—	—	3±8	0	1±7	2±9
30	—	—	—	—	—	—	—	13±12	0	0
35	—	—	—	—	—	—	—	—	0	0
40	—	—	—	—	—	—	—	—	—	0
Desert Bitterbrush Seeds										
0	0	0	0	0	0	0	0	0	0	0
2	—	0	5±4	5±3	4±3	4±3	3±3	3±3	3±4	0
5	—	—	8±4	7±2	6±2	5±2	4±2	3±2	2±3	0
10	—	—	—	11±3	9±2	7±2	5±2	3±2	1±4	0
15	—	—	—	—	11±3	8±2	5±2	2±3	0	0
20	—	—	—	—	—	9±3	5±2	0	0	0

TABLE 5.1. *(continued)*

16-Hours Cold-Period Temperature (°C)	8-Hours Warm-Period Temperature (°C)									
	0	2	5	10	15	20	25	30	35	40
25	—	—	—	—	—	—	5±3	0	0	0
30	—	—	—	—	—	—	—	2±5	0	0
35	—	—	—	—	—	—	—	—	0	0
40	—	—	—	—	—	—	—	—	—	0
Cliffrose Seeds										
0	0	0	0	0	0	0	12±5	0	3±5	0
2	—	0	0	0	0	0	13±4	12±3	5±4	0
5	—	—	0	0	8±5	13±4	14±3	13±3	8±3	1±6
10	—	—	—	2±7	7±4	13±3	15±3	15±3	11±3	4±5
15	—	—	—	—	4±5	11±3	15±3	15±3	12±3	6±5
20	—	—	—	—	—	7±4	12±3	13±3	12±4	0
25	—	—	—	—	—	—	8±4	10±3	0	0
30	—	—	—	—	—	—	—	5±4	6±3	3±5
35	—	—	—	—	—	—	—	—	0	0
40	—	—	—	—	—	—	—	—	—	0

[1]Means for temperature regimes supporting germination are underlined.

Adapted from J. A. Young and R. A. Evans, *Germination of Seeds of Antelope Bitterbrush, Desert Bitterbrush, and Cliffrose* (Agric. Res. Results 17, USDA, Agric. Res. Ser., Oakland, Calif., 1981).

not adapted for growth in salt-affected soils, so the influence of reduced osmotic potentials is not a direct measurement of normal field conditions. But it does provide an index of the influence of soil moisture stress on the cool-moist pretreatment's ability to break dormancy. Lowering the osmotic potential (the more *negative* the osmotic potential, the greater the stress on the seed) had a negative effect on the efficacy of the cool-moist pretreatment (Table 5.2).

Experiments with lowered osmotic potential involve only the potential of the solution. In actual seedbeds the *matric potential* of the soil solutions is important. The matric potential involves how tightly the films of water in the soil are attracted to the soil particles themselves and the points of contact between the soil substrate and the seed coat for hydraulic conductivity. Fine-textured substrates such as clays hold moisture much more tightly than coarse sands. The points of contact with the seed coat are greater with finer-textured materials as well. We found that as the matric potential of the substrate was made more negative, the cool-moist pretreatment was less effective in breaking dormancy (Table 5.3).

The French scientist Cone described the probable mode of action of moist prechilling in breaking the dormancy of antelope bitterbrush seeds.[8] Embryos require very little oxygen to germinate. The lower the temperature, the less oxygen is needed. After imbibition the embryo receives dissolved oxygen from water the seed coat imbibes. However, the seed coat (or fruit coat in the case of a dry achene) of many Rosaceae species contains phenolic constituents that fix part of the dissolved oxygen by oxidation and lower the quantity available to the embryo. The higher the temperature, the more oxygen the embryo requires, but the solubility of oxygen in water decreases with increasing temperature.

Recent research by Terry Booth demonstrates that the duration of moist prechilling influences the vigor of very young antelope bitterbrush seedlings.[9] If the prechilling requirements are not satisfied, the seedlings will be smaller.

Apparently, many of the early researchers who planted antelope bitterbrush seeds in field seedbeds, including Nord, concluded that two or more months of moist prechilling were required to break dormancy. Recent research we conducted in Idaho and unpublished work conducted at Doyle, California, however, clearly show that antelope bitterbrush seeds germinate and initiate root growth a few weeks after imbibition in the fall. This root elongation can reach considerable depth. In February (at Doyle) the hypocotyl arch suddenly straightens and the cotyledons pop through the soil surface. Many of the early researchers saw this early spring emergence from seeds planted the previous fall and interpreted it as evidence that antelope bitterbrush required months of field exposure to break dormancy. As a result, a lot of unnecessary research effort went into developing a chemical treatment to break the dormancy of antelope bitterbrush seeds with the hope of improving seedling establishment.

TABLE 5.2.

Germination (%) of Antelope Bitterbrush Seeds with Four Weeks' Incubation at 64°F after Cool-Moist Pretreatment

Temperature (°F)	Duration (days)	Osmotic Potential (MPa)					Untreated Seeds
		0	−0.4	−0.6	−0.8	−1.2	
36	10	52c[1]	10d–h	0h	12d–h	0h	22d
36	14	68ab	16d–g	8d–h	10d–h	4e–h	22d
43	10	56bc	20d–f	6e–h	6e–h	10d–h	22d
43	14	80a	12d–h	16d–g	18d–f	2gh	22d

[1]Means followed by the same letter are not significantly different at the 0.01 level of probability as determined by Duncan's Multiple Range Test.

Adapted from James A. Young and Raymond A. Evans, "Stratification of Bitterbrush Seeds," *J. Range Manage.* 29 (1976): 421–425.

TABLE 5.3.

Germination (%) of Antelope Bitterbrush Seeds with Four Weeks' Incubation at 64°F after Cool-Moist Pretreatment in Sand[1]

Temperature (°F)	Duration (days)	Matric Potential (MPa)					Untreated Seeds
		0	−0.4	−0.6	−0.8	−1.2	
36	10	48cd	20gh	14h	16h	20gh	22gh
36	14	72ab	14h	18gh	20gh	18gh	22gh
43	10	52cd	16h	20gh	12h	24f–h	22gh
43	14	76ab	18gh	23f–h	18gh	20gh	22gh

[1]Means followed by the same letter are not significantly different at the 0.01 level as determined by Duncan's Multiple Range Test.

Adapted from Young and Evans, "Stratification of Bitterbrush Seeds."

Researchers also failed to recognize what a great natural adaptation this winter germination and root growth followed by early spring emergence is for antelope bitterbrush. Grant Harris, for example, demonstrated the great competitive advantage winter root elongation gives to seedlings of cheatgrass (*Bromus tectorum*).[10] In the case of antelope bitterbrush, it is a double adaptation in that root elongation allows moisture extraction without exposing the tender cotyledons to frost damage and rodent or insect predation.

The stimulus that causes antelope bitterbrush seeds to emerge in the spring is unknown. There is so much variability in spring seedbed temperatures that seedbed temperature alone is unlikely to be the cause. Perhaps the number of fluctuations above a certain minimum temperature is what conditions emergence.

Robert Wagle studied the early growth variation in bitterbrush and its relation to environment for his Ph.D. dissertation at the University of California at Berkeley.[11] He used 30 accessions of *Purshia* from a wide geographic range. In testing for variation in moist-prechilling requirements he used a minimum of 6 and a maximum of 23 weeks. These excessive pretreatment periods were based on information provided by the Pacific Southwest Forest and Range Experiment Station, Forest Service, USDA. He probably got the information from Hormay, Nord, or both. The original *Woody Plant Seed Manual* published in 1948 suggests that antelope bitterbrush seeds be moist-prechilled for 2–3 months.[12] Ronald Peterson, in a study published in 1953, scarified seeds mechanically or with boiling water before moist-prechilling them—and not surprisingly got very poor germination.[13]

The myth that difficulties in artificially seeding antelope bitterbrush were due to seed dormancy was furthered by Bennett Pearson's paper "Bitterbrush Seed Dormancy Broken with Thiourea."[14] Thiourea is a very reactive chemical radical that had been known for many years to influence seed germination.[15] It is thought to interact with coumarin to induce germination, probably by mobilizing fatty acids.[16] Pearson found that soaking antelope bitterbrush seeds in a 3 percent aqueous solution of thiourea for five minutes produced germination levels of around 80 percent, results similar to those obtained using the excised embryo technique on samples from the same seed lot. Pearson saw the value of this treatment as its ability to produce germinable seeds that could be seeded in the spring, thus avoiding rodent predation over the winter and the hazards of frost. Seeds soaked in a thiourea solution and dried could be stored for later mechanical planting. This was an advantage over moist prechilling because seeds treated with that method had to be planted right away. In addition, the sticky, soft, moist seeds were very difficult for mechanical seeding devices to handle. Pearson reported that actual field results from planting thiourea-treated antelope bitterbrush seeds were erratic, but he did establish stands in which 10 percent of the planted seeds successfully made seedlings. This was at a time when the widespread seeding of crested wheatgrass (*Agropyron desertorum*) on rangelands was considered successful when 4 percent of the seeds planted resulted in established plants.

Subsequently, thiourea treatment became a near universal practice throughout the West. State fish and game departments handed out packets of treated seed to prospective hunters and even to Boy Scouts. There were three very serious errors in this widespread use of thiourea. First, there was little evidence of the biological efficacy of the treatment. Second, instructions for use of the method should have included the warning that thiourea is a highly toxic material; it is harmful to mammals in amounts of less than 50 milligrams per kilogram of body weight. Third, although no one could have known it in the 1950s when the treatment was proposed, subsequent tests determined that exposure to thiourea induced cancer in laboratory animals. Thiourea-treated seeds are certainly not something that should be given to Boy Scouts for distribution in the environment! As late as the 1980s, however, we saw a presentation on propagating antelope bitterbrush by a large commercial nursery that featured a photograph of student employees reaching barehanded into the thiourea solution to extract seed for planting in pots. Even worse, thiourea-treated seeds were sometimes additionally treated with rodent repellents before planting.

Hubbard did extensive field testing with spring-planted thiourea-treated seeds in California.[17] Seeds planted in experimental plots in Doyle, Lassen County, and Flukey Springs, Modoc County, emerged the second spring after planting. Nord was quite emphatic that antelope bitterbrush seeds would remain in the soil ungerminated despite their moist-prechilling requirements having been satisfied. The importance of this aspect of the seed physiology and ecology of antelope bitterbrush lies in the potential for building seed banks. Although we know from laboratory studies that seeds will germinate once the initial dormancy requirements are satisfied by moist chilling, we do not know if seeds rendered germinable by moist prechilling but then prevented from germinating by moisture stress will remain viable, acquire secondary dormancy, or simply die before the next annual moisture event. We have observed years when overwinter seedbeds were too dry for the initial dormancy to be broken.[18] Rodent seed predation would have to be very low, of course, for a seed bank to develop. Hubbard may have carried out his experiments in relatively rodent proof enclosures.

In collaboration with George Van Atta, Nord suggested that dormancy of antelope bitterbrush might be due to the presence of saponin in the seed (fruit) coat,[19] but as late as 1978 chemists were still looking for the dormancy material in the seeds. David Dreyer and Eugene Trousdale extracted the seeds with ethanol and hot water and got compounds with high inhibitory activity. The authors complained that these extracts contained phenolic impurities, but considering Cone's comments (noted above) about phenolic compounds and oxygen reaching the embryo, the "contaminants" may have been important.[20]

As more and more agencies began using thiourea to treat antelope bitterbrush seeds, abnormal seedlings began to appear, generating research to determine the cause.[21] Don Neal and Reed Sanderson investigated the influence of the temperature of the thiourea solution on seedling abnormalities. Normal seedling development occurred between 60 and 140°F. Below 60°F the thiourea was less effective in breaking dormancy, and above 140°F seed deformities such as cracked hypocotyls and detached root caps appeared.[22] Harper found that the longer antelope bitterbrush seeds were soaked in thiourea solution at any temperature, the greater the occurrence of abnormal seedlings.[23]

Richard Everett noted another interesting effect of the thiourea treatment: all the seeds germinated simultaneously. Everett discovered this while comparing moist-prechilled seeds with seeds whose dormancy had been broken with a completely novel method, soaking in a solution of hydrogen peroxide.[24] Untreated antelope bitterbrush seeds exhibited nearly continuous germination, and moist-prechilled seeds also emerged over a considerable period, but the thiourea-treated seeds all emerged at the same time. Everett found that this made spring-planted thiourea seeds very susceptible to frost damage.[25] In an attempt to improve seedling survival, B. R. McConnell enriched moist-prechilled antelope bitterbrush seeds with gibberellic acid.[26] He did not report any field results, but in the laboratory, gibberellic acid enrichment speeded up germination.

Hydrogen peroxide has a long history of enhancing the germination of dormant seeds. In 1916 Crocker suggested that it might increase the oxygen available to embryos.[27] Concentrated hydrogen peroxide is a powerful oxidizer. Antelope bitterbrush seeds treated with hydrogen peroxide show a visible reaction with the achene coat that results in the clearing of the dark pigments in the surface layers of the achene. Quite possibly, seed physiologists began to use hydrogen peroxide to break dormancy after first using solutions of the chemical to sterilize the seed surface.[28] Before Everett and Meeuwig discovered that hydrogen peroxide enhanced the germination of antelope bitterbrush seeds, J. W. Riffle and H. W. Springfield had determined that such treatments enhanced the germination of the related rose family species cliffrose and *Cercocarpus montanus*.[29] Everett and Meeuwig used 0.5–1 percent solutions of hydrogen peroxide with soaking times of 3–16 hours.

In our own experiments, we determined that soaking for 3 hours in a 3 percent hydrogen peroxide solution gave optimum germination with most seed sources. There was considerable variability among sources in the level of ger-

TABLE 5.4.
Mean Germination (%) Profiles for Treated and Untreated Bitterbrush and Cliffrose Seeds[1]

		Germination		
Species	Control	Moist Prechilled	Thiourea Soaked	Hydrogen Peroxide Soaked
Antelope bitterbrush	16a z	31c x	43a w	21a y
Desert bitterbrush	3c y	38a w	21c x	19b x
Cliffrose	5b x	34b w	34b w	9c x
Mean	8 y	34 w	33 w	16 x

[1] Fifty-five temperature germination profiles were used. Means in columns followed by the same letter (a–c) or means in rows followed by the same letter (w–z) are not significantly different at the 0.01 level of probability as determined by Duncan's Multiple Range Test.

Adapted from J. A. Young and R. A. Evans, *Temperature Profiles for Germination of Cool Season Range Grasses* (Agric. Res. Results 27, USDA, Agric. Res. Ser., Oakland, Calif., 1982).

mination enhancement, but all sources were enhanced to some degree by the treatment. In contrast, the response of seeds of different sources of *Cercocarpus ledifolius* soaked in hydrogen peroxide ranged from complete germination to no enhancement. The most remarkable aspect of hydrogen peroxide enhancement of the germination of antelope bitterbrush seeds is that the treated seeds can subsequently be dried, and when remoistened and placed in a suitable environment they will germinate.

We compared the germination percentages of antelope and desert bitterbrush and cliffrose seeds with no pretreatment, seeds soaked in hydrogen peroxide or thiourea solutions, and moist-prechilled seeds,[30] using 55 constant or alternating temperature regimes to produce temperature germination profiles. The germination profiles were statistically analyzed using quadratic response surfaces. The response surfaces were composed of a series of regression equations, one for each cold-period temperature through the series of warm-period temperatures with calculated values and their confidence intervals.[31] Although the information available does not begin to address the variability that must exist in *Purshia,* this is one of the few comparisons of germination parameters of the three species. Germination information of any kind is rare for desert bitterbrush and cliffrose.

Table 5.4 summarizes the differences among the three species in germination and response to dormancy-breaking treatments. When all three species were untreated before germination, antelope bitterbrush seeds had the highest mean germination (Table 5.5). Antelope bitterbrush seeds also had the highest mean germination with thiourea or hydrogen peroxide pretreatment, but both desert bitterbrush and cliffrose seeds exhibited higher mean germination for the entire profile when moist prechilling was used to break dormancy.

The unpretreated profile for antelope bitterbrush seeds shows that most of the optimum temperature regimes for germination are cold enough for self moist prechilling of the seeds with a four-week incubation period. We define optimum germination as that not lower than the maximum observed minus the confidence interval at the 0.01 level of probability. Desert bitterbrush and cliffrose seeds that were not pretreated exhibited low germination.

We tested thiourea concentrations of 0, 0.5, 1, 3, and 5 percent for durations of 1, 5, 10, 15, 30, 60, and 120 minutes. For seeds of all three species the optimum concentration and duration was 3 percent for 30 minutes—a longer soaking period than originally proposed by Pearson.[32] (Eamor Nord suggested the longer soaking period to Young.) We did not observe seedling abnormalities with this longer soaking. It is apparent from Table 5.6 and the profiles themselves (Table 5.7) that the thiourea treatment greatly enhanced germination of all three species. Desert bitterbrush was the least enhanced, and antelope bitterbrush seeds showed some germination at all temperature regimes except the extremes. The breadth of temperature regimes at which some germination occurred increased to 85–95 percent of the 55 regimes tested, and overall mean germination climbed from 3–16 percent to 21–43 percent. Thus, thiourea treatment not only increased total germination, it also broadened the temperature extremes at which some germination would occur.

Each of the three species showed a distinct response to moist-prechilling temperatures in subsequent germination. We present data for the 43°F (5°C) regime only, but data for all the temperatures are available in the original publication.[33] The warmest moist-prechilling temperature (43°) reduced the required time to break dormancy to two weeks for all species. The germination profiles for all three species were similar to those obtained with thiourea (Tables 5.8 and 5.9).

The enhancement of germination was less with hydrogen peroxide pretreatment than with thiourea in all three species. Cliffrose seeds were the least enhanced (Tables 5.10 and 5.11).

When a relatively large number of temperature-germination profiles are produced, as was done in our studies of the three species of *Purshia,* it is possible to calculate meaningful frequencies of temperature regimes that support

TABLE 5.5.
Germination Profiles (%) for Bitterbrush and Cliffrose Seeds without Dormancy-Breaking Treatments before Incubation[1]

Parameter	Species		
	Cliffrose	Desert Bitterbrush	Antelope Bitterbrush
Regimes with some germination	60	51	87
Mean germination	5	3	16
Maximum germination	15	32	11
Regimes with optimum germination	27	11	22
Mean of optima	13	8	30

[1] Seeds were incubated for four weeks at 55 constant and alternating temperatures.

Adapted from Young and Evans, *Temperature Profiles for Germination.*

TABLE 5.6.
Germination Profiles (%) for Bitterbrush and Cliffrose Seeds after Soaking for 30 Minutes in a 3 Percent Thiourea Solution[1]

Parameter	Species		
	Cliffrose	Desert Bitterbrush	Antelope Bitterbrush
Regimes with some germination	93	85	95
Mean germination	34	21	43
Maximum germination	62	46	67
Regimes with optimum germination	16	15	18
Mean of optima	66	43	65

[1] Seeds were incubated for four weeks at 55 constant and alternating tempertures.

Adapted from Young and Evans, *Temperature Profiles for Germination.*

TABLE 5.7.

Estimated Percentage Germination of Antelope and Desert Bitterbrush and Cliffrose Seeds, Soaked in 3 Percent Solution of Thiourea for 30 Minutes Before Incubation[1]

16-Hours Cold-Period Temperature (°C)	8-Hours Warm-Period Temperature (°C) 0	2	5	10	15	20	25	30	35	40
Antelope Bitterbrush Seeds										
0	0	50±10	55±8	60±6	61±7	59±7	53±7	43±7	29±8	12±11
2	—	51±9	56±7	62±5	64±5	63±6	57±6	48±5	36±6	19±9
5	—	—	57±8	64±5	67±4	66±5	62±5	54±5	43±5	28±8
10	—	—	—	60±7	65±5	67±5	65±5	59±5	49±5	36±7
15	—	—	—	—	57±8	60±6	60±5	56±5	48±6	37±8
20	—	—	—	—	—	47±8	48±6	46±5	41±6	31±8
25	—	—	—	—	—	—	30±8	30±5	26±6	18±8
30	—	—	—	—	—	—	—	6±9	4±8	1±10
35	—	—	—	—	—	—	—	—	25±13	0
40	—	—	—	—	—	—	—	—	—	0
Desert Bitterbrush Seeds										
0	0	13±8	3±7	10±6	19±6	24±6	24±6	20±6	11±7	2±11
2	—	9±8	1±6	15±5	24±5	29±5	30±5	26±5	18±6	0
5	—	—	5±7	20±4	30±4	36±4	37±4	34±4	26±5	14±8
10	—	—	—	24±7	35±5	42±5	44±5	42±5	35±6	0
15	—	—	—	—	35±7	42±5	46±5	45±5	39±6	39±9

TABLE 5.7. *(continued)*

16-Hours Cold-Period Temperature (°C)	8-Hours Warm-Period Temperature (°C)									
	0	2	5	10	15	20	25	30	35	40
20	—	—	—	—	—	38±7	42±5	42±5	37±7	0
25	—	—	—	—	—	—	33±8	34±6	30±7	0
30	—	—	—	—	—	—	—	20±10	18±10	0
35	—	—	—	—	—	—	—	—	0	0
40	—	—	—	—	—	—	—	—	—	0
Cliffrose Seeds										
0	—	0	18±11	36±9	48±8	53±9	53±8	46±8	34±9	15±14
2	—	8±13	21±10	39±7	51±7	57±7	57±7	50±7	38±8	20±12
5	—	—	24±11	42±6	54±6	60±6	60±6	54±6	42±6	25±10
10	—	—	—	42±9	54±7	61±6	62±6	57±6	45±6	28±9
15	—	—	—	—	49±10	56±7	57±7	53±7	42±7	25±10
20	—	—	—	—	—	45±10	47±7	43±6	33±7	17±10
25	—	—	—	—	—	—	31±10	27±7	18±7	2±10
30	—	—	—	—	—	—	—	6±11	3±10	18±12
35	—	—	—	—	—	—	—	—	30±16	0
40	—	—	—	—	—	—	—	—	—	0

[1] Means for temperature regimes supporting germination are underlined.

Adapted from Young and Evans, *Germination of Seeds.*

TABLE 5.8.
Germination Profiles (%) for Bitterbrush and Cliffrose Seeds after Two Weeks' Moist Prechilling at 36°F (2°C)[1]

Parameter	Cliffrose	Desert Bitterbrush	Antelope Bitterbrush
	Species		
Regimes with some germination	96	80	91
Mean germination	34	38	31
Maximum germination	58	77	54
Regimes with optimum germination	15	11	23
Mean of optima	55	72	52

[1]Seeds were incubated for four weeks at 55 constant and alternating temperatures.

Adapted from Young and Evans, *Temperature Profiles for Germination.*

optimum germination (Table 5.12). For antelope bitterbrush, the optima were all at cold-period temperatures from 32 through 64°F (0–15°C) alternating with 53–96°F (10–30°C) warm periods. The only temperature regime that always supported optimum germination was 43/74°F (5/20°C). The optimal temperature regimes for desert bitterbrush were slightly warmer, with the exception of a low frequency of germination at a constant 43°F (5°C) (see Table 5.12). There was no 100 percent frequency optimum for desert bitterbrush. The highest frequency optima occurred at 53 and 64°F (10 and 15°C) alternating with 74°F (20°C).

The optimal temperature regimes for cliffrose spanned a wide range of temperatures (Fig. 5.2) from a low of a constant 53° (10°C) to a high of 74/106°F (20/25°C). There were four temperatures with a 100 percent frequency of optima clustered near 53/85°F (10/25°C).

The real importance of germination-temperature profiles is in comparing the response of the seeds to seedbed temperatures monitored in the field. Raymond Evans pioneered research in electronically monitoring seedbed temperatures during the germination period in sagebrush rangeland communities.[34] Using the data from field monitoring we divided the temperature germination profile regimes into a grouping of seedbed temperatures (Fig. 5.2). At what we class as moderate seedbed temperatures any of the dormancy-breaking treatments at least doubled the mean germination of antelope bitterbrush seeds (Table 5.13).

TABLE 5.9.

Estimated Percentage Germination of Antelope and Desert Bitterbrush and Cliffrose Seeds, Moist-Prechilled for Two Weeks at 43°F[1]

16-Hours Cold-Period Temperature (°C)	8-Hours Warm-Period Temperature (°C)									
	0	2	5	10	15	20	25	30	35	40
Antelope Bitterbrush Seeds										
0	—	38±10	43±8	48±7	51±7	50±7	46±7	34±7	30±8	17±11
2	—	40±10	44±7	50±5	52±6	52±6	48±6	51±5	32±6	19±10
5	—	—	46±8	51±5	54±5	53±5	50±5	43±5	33±5	20±8
10	—	—	—	50±8	53±6	52±5	48±5	42±5	32±6	19±8
15	—	—	—	—	47±9	46±6	43±6	36±6	27±6	14±8
20	—	—	—	—	—	36±8	33±6	26±6	17±6	4±9
25	—	—	—	—	—	—	18±9	12±7	2±7	11±10
30	—	—	—	—	—	—	—	8±12	17±11	0
35	—	—	—	—	—	—	—	—	0	0
40	—	—	—	—	—	—	—	—	—	0
Desert Bitterbrush Seeds										
0	0	34±12	38±10	43±9	45±9	43±9	37±9	38±10	15±15	0
2	—	42±12	46±9	51±7	52±7	50±8	44±7	34±8	21±12	0
5	—	—	57±10	61±6	61±6	58±7	52±6	42±6	28±9	0
10	—	—	—	72±10	72±7	68±7	61±7	50±7	36±9	0
15	—	—	—	—	77±11	72±8	64±7	52±7	37±9	18±14

TABLE 5.9. *(continued)*

16-Hours Cold-Period Temperature (°C)	8-Hours Warm-Period Temperature (°C)									
	0	2	5	10	15	20	25	30	35	40
20	—	—	—	—	=	71±11	61±7	49±7	33±10	0
25	—	—	—	—	—	—	53±11	40±8	23±11	0
30	—	—	—	—	—	—	—	25±15	7±15	0
35	—	—	—	—	—	—	—	—	0	0
40	—	—	—	—	—	—	—	—	—	0
Cliffrose Seeds										
0	0	0	14±8	32±7	44±7	50±7	50±7	45±7	34±8	17±12
2	—	2±10	16±8	34±6	46±6	52±6	53±6	48±6	37±7	20±10
5	—	—	18±8	36±5	48±5	55±5	56±5	51±5	40±5	24±8
10	—	—	—	36±8	49±5	56±5	58±5	53±5	43±5	27±8
15	—	—	—	—	47±8	55±6	56±5	52±5	42±6	27±8
20	—	—	—	—	—	44±9	51±5	48±5	38±6	23±9
25	—	—	—	—	—	—	43±8	40±6	31±6	16±8
30	—	—	—	—	—	—	—	28±8	20±6	5±8
35	—	—	—	—	—	—	—	—	5±10	9±10
40	—	—	—	—	—	—	—	—	—	26±14

[1]Means for temperature regimes supporting germination are underlined.

Adapted from Young and Evans, *Germination of Seeds*.

TABLE 5.10.
Germination Profiles (%) for Bitterbrush and Cliffrose Seeds after Six Hours' Soaking in a 1 Percent Hydrogen Peroxide Solution[1]

Parameter	Species		
	Cliffrose	Desert Bitterbrush	Antelope Bitterbrush
Regimes with some germination	62	87	85
Mean germination	9	19	21
Maximum germination	28	39	45
Regimes with optimum germination	20	13	15
Mean of optima	25	37	49

[1] Seeds were incubated for four weeks at 55 constant and alternating temperatures.

Adapted from Young and Evans, *Temperature Profiles for Germination.*

The percentage increase was even greater for desert bitterbrush and cliffrose (except for hydrogen peroxide–treated cliffrose seeds). Antelope bitterbrush seeds demonstrated comparatively good germination at colder seedbed temperatures. In the case of desert bitterbrush, only moist-prechilled seeds showed good germination at colder seedbed temperatures. Cliffrose controls did not germinate at colder seedbed temperatures, but moist-prechilled or thiourea-treated seeds did germinate. Germination of all three species declined at warmer seedbed temperatures, with the apparent paradox that moist-prechilled cliffrose seeds germinated better than control seeds at warmer temperatures. Fluctuating seedbed temperatures are important because rodents cache seeds of these shrubs fairly close to the soil surface. Antelope bitterbrush seeds demonstrated good germination in all categories with this type of seedbed temperature regime. And for desert bitterbrush, the dormancy-breaking pretreatments greatly increased germination at widely fluctuating temperatures. Obviously, a lot of research time and money has been devoted to studying the seed and seedbed ecology of *Purshia* species, and a lot is known about the factors critical to seed germination. The most meaningful seed physiology information is that which is directly related to how the seeds function in field seedbeds.

TABLE 5.11.

Estimated Percent Germination for Antelope and Desert Bitterbrush and Cliffrose Seeds, after 6 Hours' Soaking in 3 Percent Hydrogen Peroxide[1]

16-Hours Cold-Period Temperature (°C)	8-Hours Warm-Period Temperature (°C)									
	0	2	5	10	15	20	25	30	35	40
Antelope Bitterbrush Seeds										
0	—	13±8	19±6	27±5	32±5	32±5	29±5	22±5	12±6	3±8
2	—	18±7	24±5	32±4	36±4	37±4	33±4	26±4	16±5	1±7
5	—	—	30±6	38±4	42±4	42±4	38±4	31±4	20±4	5±6
10	—	—	—	42±6	45±5	45±4	41±4	33±4	22±4	7±6
15	—	—	—	—	43±7	42±4	38±4	30±4	18±5	3±7
20	—	—	—	—	—	34±6	29±4	20±4	8±6	8±8
25	—	—	—	—	—	—	13±7	5±6	0	0
30	—	—	—	—	—	—	—	0	0	0
35	—	—	—	—	—	—	—	—	0	0
40	—	—	—	—	—	—	—	—	—	0
Desert Bitterbrush Seeds										
0	0	1±6	11±8	12±5	21±5	26±6	26±5	21±5	13±6	1±9
2	—	2±6	9±8	15±4	25±4	30±5	30±5	26±4	18±5	5±8
5	—	—	4±6	19±4	29±4	34±4	35±4	32±4	24±4	11±7
10	—	—	—	21±6	31±4	38±4	39±4	37±4	30±5	21±8
15	—	—	—	—	29±7	36±5	39±4	38±5	31±6	0

TABLE 5.11. *(continued)*

16-Hours Cold-Period Temperature (°C)	8-Hours Warm-Period Temperature (°C)									
	0	2	5	10	15	20	25	30	35	40
20	—	—	—	—	—	31±7	34±5	33±4	28±6	0
25	—	—	—	—	—	—	25±6	25±5	20±6	0
30	—	—	—	—	—	—	—	11±7	8±7	0
35	—	—	—	—	—	—	—	—	10±11	0
40	—	—	—	—	—	—	—	—	—	0
Cliffrose Seeds										
0	—	0	0	0	0	4±7	3±7	0	0	0
2	—	0	1±13	6±8	9±6	10±6	9±6	6±6	0	0
5	—	—	0	0	17±5	18±5	17±5	13±5	8±7	1±12
10	—	—	—	23±9	25±6	26±5	24±5	21±5	16±6	8±10
15	—	—	—	—	28±8	28±6	26±5	22±6	17±7	0
20	—	—	—	—	—	25±8	23±5	19±6	13±8	0
25	—	—	—	—	—	—	14±8	9±6	0	0
30	—	—	—	—	—	—	—	5±11	12±12	0
35	—	—	—	—	—	—	—	—	0	
40	—	—	—	—	—	—	—	—	—	0

[1]Means for temperature regimes supporting germination are underlined.

Adapted from Young and Evans, *Germination of Seeds.*

TABLE 5.12.

Frequency of Optimum Temperature Regimes for Germination of Antelope and Desert Bitterbrush and Cliffrose Seeds, Incubated at 55 Constant or Alternating Temperature Regimes with Treatments of Control, Moist Prechilling, Thiourea, and Hydrogen Peroxide

16-Hours Cold-Period Temperature (°C)	8-Hours Warm-Period Temperature (°C)									
	0	2	5	10	15	20	25	30	35	40
Antelope Bitterbrush Seeds										
0	—	—	—	—	50	25	25	25	—	—
2	—	—	—	25	75	75	25	25	—	—
5	—	—	—	25	75	100	25	25	—	—
10	—	—	—	25	75	75	75	—	—	—
15	—	—	—	—	25	25	—	—	—	—
20	—	—	—	—	—	—	—	—	—	—
25	—	—	—	—	—	—	—	—	—	—
30	—	—	—	—	—	—	—	—	—	—
35	—	—	—	—	—	—	—	—	—	—
40	—	—	—	—	—	—	—	—	—	—
Desert Bitterbrush Seeds										
0	0	—	—	—	—	—	—	—	—	—
2	—	—	—	—	—	—	—	—	—	—
5	—	—	25	—	—	—	25	—	—	—
10	—	—	—	50	50	75	50	50	—	—

TABLE 5.12. *(continued)*

16-Hours Cold-Period Temperature (°C)	8-Hours Warm-Period Temperature (°C)									
	0	2	5	10	15	20	25	30	35	40
15	—	—	—	—	50	75	50	50	—	—
20	—	—	—	—	—	50	25	25	—	—
25	—	—	—	—	—	—	—	—	—	—
30	—	—	—	—	—	—	—	—	—	—
35	—	—	—	—	—	—	—	—	—	—
40	—	—	—	—	—	—	—	—	—	—
Cliffrose Seeds										
0	—	—	—	—	—	—	25	—	—	—
2	—	—	—	—	—	25	75	25	—	—
5	—	—	—	—	—	75	75	25	—	—
10	—	—	—	25	25	100	100	100	—	—
15	—	—	—	—	25	75	100	50	—	—
20	—	—	—	—	—	25	50	20	25	—
25	—	—	—	—	—	—	—	—	—	—
30	—	—	—	—	—	—	—	—	—	—
35	—	—	—	—	—	—	—	—	—	—
45	—	—	—	—	—	—	—	—	—	—

Adapted from Young and Evans, *Germination of Seeds.*

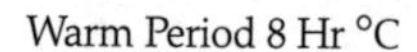

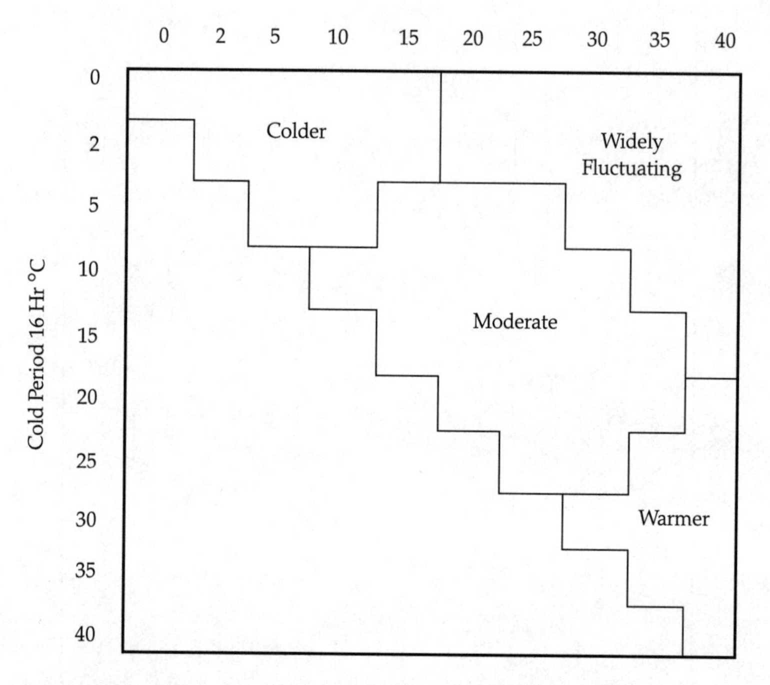

Fig. 5.2. Categories of seedbed temperatures used for comparison with germination profiles. Adapted from Young and Evans, *Germination of Seeds*.

TABLE 5.13.
Mean Germination (%) for Four Seedbed Temperature Regimes and Treatments

Species and Treatment	Categories of Seedbed Temperatures			
	Moderate	Colder	Warmer	Widely Fluctuating
Antelope bitterbrush				
Control	16	17	3	25
Moist prechilled	39	42	4	33
Thiourea	54	58	12	42
Hydrogen peroxide	32	25	1	19
Desert bitterbrush				
Control	4	2	—	1
Moist prechilled	58	43	6	25
Thiourea	38	11	8	20
Hydrogen peroxide	32	11	5	20
Cliffrose				
Control	11	0	2	5
Moist prechilled	49	22	18	39
Thiourea	49	26	10	40
Hydrogen peroxide	21	2	2	5

Adapted from Young and Evans, *Temperature Profiles for Germination.*

Chapter Six

Seeding *Purshia* Species

The very first publications concerning antelope bitterbrush ecology made it clear that natural populations, especially those in critical mule deer wintering areas, were not renewing themselves and that artificial establishment of the valuable shrub appeared to be highly desirable. In his pioneer treatment of antelope bitterbrush August Hormay touched on most of the appropriate topics concerning the artificial seeding of this species. A tremendous amount of research has since been directed toward refining the process, but the categories used by Hormay still apply:[1]

1. Collecting, processing, and storing seed
2. Selection of planting site
3. Time of planting
4. Preplanting seed treatments
5. Depth and method of planting and rate
6. Seedling competition
7. Rodent predation

Hormay reported that seeds on the lower branches matured first and suggested collecting mature seeds by using a gloved hand to strip them from the branches. Determining the maturity of seeds is not difficult. Antelope bitterbrush seeds become heavily pigmented with a dark red sap before maturity. If you crush an immature seed between your thumb and forefinger the pigment stains your skin. Many a range conservationist or wildlife manager has initiated a summer assistant by convincing him or her to squash the immature seed and then taste the red sap. The red stain is indelible, and once tasted, the name bitterbrush will never be forgotten. When the seed is too indurate to crush and the red pigment has disappeared (Hormay suggested that it crystallized in the achene coat), the seeds are ripe. Hormay also noted the briefness of the harvest period. A stiff wind can disperse the entire seed crop in a single day.

Eamor Nord made the harvesting of antelope bitterbrush seed a much more scientific endeavor, first in an article titled "Bitterbrush Seed Harvesting: When, Where, and How," and later in his monograph on bitterbrush.[2] He reinforced Hormay's observation that the seeds must be harvested quickly once they are ripe, before they disperse. Nord also noted that virtually all the plants

Fig. 6.1. Screen tray and paddle for harvesting antelope bitterbrush seed. From E. C. Nord, "Bitterbrush Seed Harvesting: When, Where, and How," *J. Range Manage.* 16 (1963): 258–261.

in a given stand matured at the same time, and he was the first to use screen-covered panels to harvest seeds. Each panel was about a yard square with a central handle and a 2-inch-high outside lip. Collectors held the panel under the shrub with one hand and beat the seed from the plant with a wooden paddle held in the other (Fig. 6.1).[3] Nord claimed that an energetic worker in a good stand could harvest 50 pounds of seed per day (30–35 pounds of clean seed); average workers could collect a pound of seed per hour. Nord also reported on a vehicle-mounted, power-driven vacuum harvester that was successfully used to harvest bitterbrush seed (Fig. 6.2). A three-man crew using the vacuum harvester (one driving and two harvesting) could harvest 160 pounds of bitterbrush seed per day.

In 1963 the Forest Service took advantage of an exceptional antelope and desert bitterbrush seed production year to harvest more than 13,700 pounds of seed, which yielded 5,000 pounds of clean seed (1 pound for each 2.74 pounds

Fig. 6.2. Vacuum browse-seed harvester under development by the USDA, Forest Service, in the 1960s. From Nord, "Bitterbrush Seed Harvesting."

collected).[4] This abundance provided a chance to test the browse-seed collector developed by the San Dimas Equipment Center, Forest Service, USDA. This vacuum-suction machine, mounted on a four-wheel-drive vehicle, sucked in seeds via two 20-foot-long hoses supported by swing booms mounted on the vehicle. The reported yield, 1 pound of clean seeds per 3.57 pounds of field-collected seeds, indicates that a considerable amount of leaves and trash were picked up along with the seeds.

The Inyo seed tray, a device that originated in the eastern Sierra, offered another collection option. The tray was a wooden-framed, round-bottomed structure, 20 inches long and 30 inches wide with a wooden rod down the center that projected about a foot from the end to serve as a handle. The locals thought it was much easier to use than larger aluminum or screen trays. In Utah, bitterbrush seed collectors used a hoop with a cloth bag attached (Fig. 6.3).[5] Commercial collectors improvised by using cardboard boxes.

Seed storage is best accomplished in a cool, dry place, preferably in tins safe

Fig. 6.3. Hoop and cloth bag for bitterbrush seed collection. From B. C. Giunta, R. Stevens, K. R. Jorgensen, and A. P. Plummer, *Antelope Bitterbrush — an Important Wildland Shrub* (Publ. 78–12, Utah Div. Wildl. Res., Salt Lake City, 1978).

from rodents, which are attracted by unprotected seeds. Desert and antelope bitterbrush and cliffrose seeds have been stored for as long as 25 years under warehouse conditions while still maintaining better than 60 percent germination.[6] Before placing the seeds in closed containers, several basic points of seed handling and storage should be considered.

At the time of collection, antelope bitterbrush seeds have a moisture content too high for storage.[7] They need to be given the opportunity to reach equilibrium with atmospheric moisture. Field-collected material should be rough-screened to remove twigs, leaves, and other material. This both reduces the moisture content and helps to remove seed-eating insects. The seeds should be spread in a dry area. If they are dried outdoors they must be covered with a moisture-proof tarp at night. In the far western United States, the relative humidity during the summer is usually sufficiently low for seeds to reach moisture equilibrium without extra measures being required.

Seeds should be dried to about 14 percent moisture and then maintained under low-moisture conditions. All nonrecalcitrant seeds (i.e., seeds that do not either germinate or become nonviable immediately after maturation) are initially dormant because of their lowered moisture content. If the moisture content rises to about 30 percent, nonsecondary dormant seeds (i.e., seeds that develop dormancy after they are dispersed from the mother plant) will try to germinate. In addition, microorganisms will grow on moist stored seeds. Keeping the seeds in shallow piles and turning them frequently aids in uniform drying. Artificial drying with heat is not a simple procedure and can easily damage the seeds. Noncoated paper, burlap, and woven synthetic fiber bags are excellent for initial seed storage because they allow for further moisture equilibration. Never store seeds, especially freshly harvested seeds, in plastic bags, not even when transporting them from the field to a processing center. A black plastic bag in an open truck on a summer day is an excellent seed steamer.

Bitterbrush seeds require special care in the threshing process. When dry, the achene is quite indurate, but the germ (the pointed end of the achene containing the embryonic plants) can be damaged in careless handling. A quick examination of any seed lot of bitterbrush is usually enough to detect excessive threshing damage. Most commercial seed shows some rubbing damage on the thickest portion of the achene, but an occasional rub-damaged seed is not going to significantly lower the value of the seed lot. If you find achenes with terminal damage, sample the seed lot in more detail to determine the extent of the damage. Nord et al. suggested running field-collected bitterbrush seeds through a hammer mill as an initial step in processing,[8] but we consider this to be very damaging to seed quality. Most commercial operators use rubber-belt threshers with adjustable clearance on the belts for the initial threshing. In Utah, seed collectors use a barley debearder to knock the papery style off the achene (Fig. 6.4) and an air screen for the final cleaning (Fig. 6.5). Many older publications refer to the threshing process as "removing the seed from the fruit." The remnant style is really an attending flower part, however, and the achene is a dry fruit.

Most lots of antelope bitterbrush seed will contain black, shriveled seeds that have been damaged by frost, insects, or drought (Fig. 6.6). R. B. Ferguson noted a different kind of seed damage that he called spotted seed. The germination of spotted seeds was only 50–80 percent of that of nonspotted seeds. Ferguson thought the spots were caused by insects, but he was not able to confirm that theory.[9]

Hormay recommended checking the viability of freshly harvested seeds by soaking them in water, cutting the achene coat with a razor blade, and dissecting out the embryo. The embryo should appear bright white and not shriveled or dark. Nord later refined this method into a reliable viability test (see Chapter 5).

Fig. 6.4. Barley debearder for removing remnant style parts from antelope bitterbrush achenes. From Giunta et al., *Antelope Bitterbrush.*

The question of where to plant bitterbrush would seem to have a simple answer. If you are interested in creating more browse for mule deer, you should plant it everywhere. Closer analysis, however, would suggest planting it only where it is adapted to grow. In Chapter 3 we described natural antelope bitterbrush communities and pointed out that natural regeneration of bitterbrush is not a problem in most ponderosa–Jeffrey pine/antelope bitterbrush woodlands. The problem sites for natural regeneration of antelope bitterbrush stands are the lower elevation sites that are critical winter range for mule deer.

Land managers often want to reseed burned winter range areas with antelope bitterbrush. The revegetation is usually undertaken as "emergency restoration of game range," but if you consider that it takes about 10 years for an antelope bitterbrush stand to reach the point at which it can sustain browsing, the use of the word *emergency* is somewhat misleading.

Fig. 6.5. Air screen for cleaning antelope bitterbrush seed. From Giunta et al., *Antelope Bitterbrush.*

Several interacting factors must be considered when artificially seeding antelope bitterbrush on burned areas. First, if a moderately dense to dense stand of antelope bitterbrush and big sagebrush is consumed in a wildfire, the seedbed will be relatively free of competition from cheatgrass during the first growing season after the fire. That is, the window is open for seedling establishment for one year. After that, cheatgrass will begin to dominate the site, taking advantage of nutrients and moisture released by the destruction of the previous vegetation.[10] Once cheatgrass becomes the dominant species, reburning will not open the potential revegetation window because there will not be enough woody fuel to reduce the cheatgrass seed banks (see Chapter 10). The briefness of this window does create an "emergency" aspect to seeding critical mule deer winter range.

Habitat managers must also realize that the nutritious sprouts and seedlings that spring up in burned areas attract browsing animals, a point Nord made in his monograph on antelope bitterbrush. The seedlings must be protected if they are to have a chance. Domestic livestock can be excluded from reseeded burns by fencing or closing existing allotments, but mule deer can be excluded only

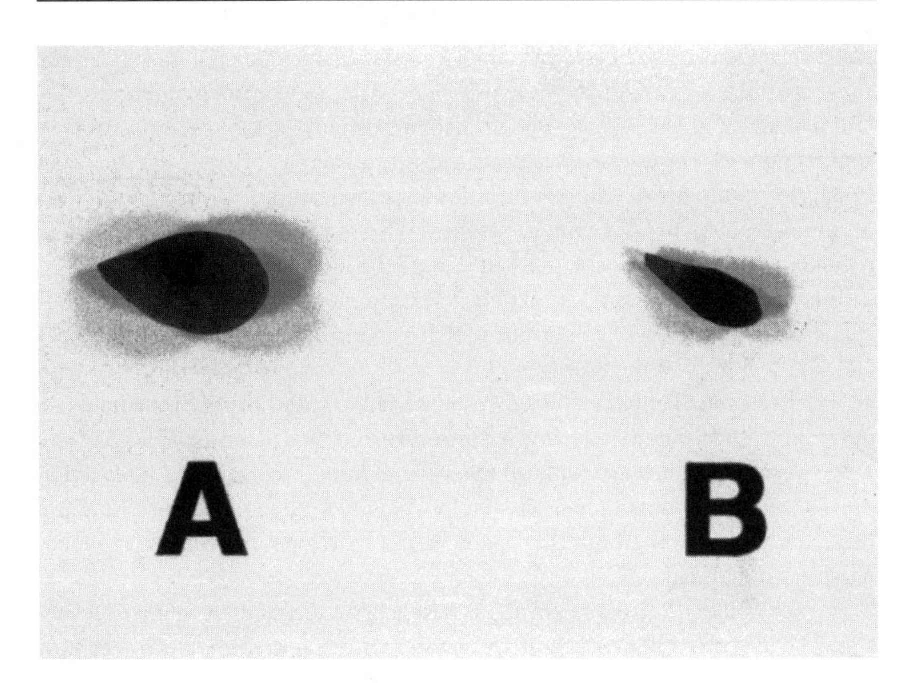

Fig. 6.6. Normal (a) and black shriveled (b) antelope bitterbrush seeds. The black seeds may have been damaged by frost, insects, or drought.

by hunting or by very expensive mule deer–proof fencing. At times it may be necessary to fence the area to exclude jackrabbits. T. A. Phillips described an antelope bitterbrush seeding in Idaho that was completely lost to jackrabbits.[11]

Another important consideration in burn rehabilitation is the suppression of subsequent wildfires. On the eastern flank of the Sierra Nevada the frequency of wildfire reoccurrence is less than 10 years in the mule deer winter range areas. We have just suggested that it takes 10 years for antelope bitterbrush seedlings to reach a size sufficient to sustain browsing. Obviously, then, in such a situation it is just as important to seed fuel breaks for wildfire management as it is to seed antelope bitterbrush. For many years wildlife managers reseeded with antelope bitterbrush alone rather than designing planting strategies that used additional types of plants to suppress wildfire. If they succeeded in establishing shrubs alone, cheatgrass invaded the stand, and eventually the shrub stand was lost to wildfire. Only perennial grasses can suppress cheatgrass, but the most successful of these, crested wheatgrass, a common cattle forage, competes with bitterbrush seedlings. There are ways to grow both in the same area, and we discuss these later in this chapter. The critical factor is to plan a plant community that biologically suppresses wildfires.

Some state game management agencies refuse to use crested wheatgrass to reseed burned big sagebrush communities because it is not a native species and they face political pressure from native plant societies to avoid planting exotics. Unfortunately, there seems to be no native perennial grass species capable of competing with cheatgrass. Attempts to restore habitat for mule deer in sagebrush environments that do not include crested wheatgrass in the planting mix are almost certain to fall victim to wildfires. This is a basic biological reality, and no amount of political posturing will change the ultimate outcome.

Too often, it seems, sites for artificial seeding of antelope bitterbrush are selected following the line of reasoning that antelope bitterbrush is a good species, so it belongs on the best sites! Every scientist who has experimented with artificial seeding of antelope bitterbrush has had to suffer through the field trip question, "How come you can't grow bitterbrush in your plots on this good site when we saw it on raw road cuts all the way out here?" We offer an explanation of this well-documented phenomenon in Chapter 8 when we discuss soils and nitrogen, but here the important point is that antelope bitterbrush colonizes harsh, *open* habitats.

Russian thistle (*Salsola australis*) is perhaps the ultimate in colonizing species. The superabundant seeds produced by each plant are relatively cheap for the mother plant to make. Essentially, each seed is a tiny embryonic plant with a minimum of covering. The tumbling action of mature Russian thistle assures widespread dispersal of the seeds.[12] Although Russian thistle is an annual and antelope bitterbrush is a woody perennial with a potential longevity greater than 100 years, seedlings of both species can exploit harsh, nutrient-poor seedbeds. Neither species can successfully establish in the face of severe competition from other plant species. As a perennial, however, antelope bitterbrush has several advantages over Russian thistle. It is not obligated to annual mass seedling establishment, as Russian thistle is. And being a long-lived perennial, antelope bitterbrush can dominate a site it colonizes for many years. Mass antelope bitterbrush establishment may be a rare event in nature, but the consequences of that event can be evident for a century. Unfortunately, the best site in terms of parameters for plant growth and productivity is often the site with the greatest competition for seedling establishment.

Game departments tend to think big when they plan antelope bitterbrush seeding projects. Give every Boy Scout a package of thiourea-treated seeds for random planting; collect seed, grow seedlings in nurseries, and set out 80,000 bare-root transplants. In terms of selecting sites for artificially seeding bitterbrush, however, it does not hurt to think small. We put this philosophy into practice in revegetating the Balls Canyon wildfire north of Reno, Nevada. The site is located on the eastern slope of the Sierra Nevada just below the ecotone between mountain big sagebrush and ponderosa–Jeffrey pine woodlands. The sites generally had a mountain big sagebrush overstory with cheatgrass dominating under the shrubs. There were occasional heavily utilized, decadent antelope bitterbrush plants in several different communities in the area before it burned. The canyon is a major transitional area for mule deer moving from summer to winter ranges and is considered critical habitat.

We found a multitude of microsites within the burn, including steep gulches, stringer meadows, alluvial fans of various ages and textures, and residual soils on steep slopes. We used a 40 horsepower wheel tractor and a half-sized rangeland drill to seed the lower portions of the burn. The seeding mixture was tailored for each microsite, with consideration given to providing escape cover for mule deer as well as browse. Antelope bitterbrush was seeded by itself and in combination with grasses and forbs on different sites. The antelope bitterbrush seedlings were wrapped inside perennial grass seedlings for biological suppression of wildfires. We believe this concept has great potential in artificial antelope bitterbrush seedings.

Land managers throughout the Intermountain states are always faced with the decision of when to seed: in the fall after the summer drought, or the winter with its heavy precipitation? Joseph H. Robertson used to claim that people who argued about which season was best were wasting their time because the fall seeders seeded so late in the fall that it was really early winter seeding, and the spring seeders tried to seed so early in the spring that it was really late winter seeding. Why not call both winter seeding, he asked, and quit arguing? There have always been a few scientists who have insisted that spring seeding is the answer to artificial regeneration. The problem with spring seeding is that the dormancy of the seeds must be overcome artificially by either moist prechilling or treating them with thiourea or hydrogen peroxide (see Chapter 5). The only valid arguments for spring seeding are less rodent predation and, on specific soils, less frost heaving.

Richard Hubbard, a proponent of spring planting, reported on several trials in northeastern California, but so many interacting factors affected his plots (e.g., periodicity and amount of soil moisture, soil temperature, frost, insects, soil-borne diseases, and rodent predation) that it was impossible to say clearly which was best in a given year.[13] Hubbard had the best success with spring seeding at higher elevation sites in Modoc County, California, in western juniper woodlands and the margins of ponderosa–Jeffrey pine woodlands. At Doyle, California, in a big sagebrush environment, spring seeding was clearly not advantageous.

For reasons discussed in Chapter 5, thiourea treatment is no longer an acceptable method for breaking seed dormancy. Instead, the seeds must be either moist-prechilled and planted wet or treated with hydrogen peroxide. Hydrogen peroxide pretreatment has not been tested on a scale sufficiently large to assess results. We suggest the following procedure for treating field-sized quantities

(30 pounds) of antelope bitterbrush seeds with hydrogen peroxide to break dormancy:

Step 1. Divide 30 pounds of clean, threshed antelope bitterbrush seed into ten 3-pound lots.

Step 2. Spread seeds in flat trays that measure about 12 by 14 by 2 inches. The dimensions do not have to be exact; any similar-sized tray will do. Pour 3 pounds of seed into each tray.

Step 3. Wearing skin and eye protection, add about 3 quarts of 3 percent hydrogen peroxide solution to each tray. Stir until all the seeds are thoroughly covered. A 3 percent hydrogen peroxide solution is relatively safe to handle. Stronger concentrations are very dangerous both because it is a highly reactive material and because it will painlessly remove skin without your being aware of it.

Step 4. Let seeds soak for 5 hours, stirring three times during that period.

Step 5. Recover seeds by screening and rinse them with tap water. Let them drain for 2 hours and then spread on paper to dry in a warm, dry place for 72 hours. Stir occasionally for even drying.

Step 6. Treated seeds can be rebagged for storage or transportation to the field. Label the seeds to indicate that they have been treated with hydrogen peroxide.

The fact that seedlings begin significant root growth as soon as the moist prechilling requirements are met, even though no shoot growth was visible beforehand, certainly appears to be a big advantage for fall seeding.

The rate and method of planting antelope bitterbrush seeds have been the subjects of as much controversy as the time of planting. Naturally established antelope bitterbrush plants often appear to grow in tight groups that arose from the same rodent seed cache. This fact has led some researchers to assume that groups of seedlings might have some advantage over single seedlings. The major form of competition in seedbeds is for soil moisture, as Raymond Evans showed in studies using microenvironmental monitoring techniques.[14] Thus, grouping seedlings should contribute to intraspecific competition for moisture and outweigh any benefits clustering might have. This is not the case. In a study conducted in the Boise River watershed in Idaho, R. B. Ferguson and Joseph Basile found a significant correlation between the number of antelope bitterbrush seedlings emerging per cluster and successful seedling establishment.[15] Their technique was based on the research of Ralph Holmgren, who studied competition between antelope bitterbrush seedlings and cheatgrass.[16] To reduce competition from other species, the vegetation and surface soil were peeled off a square or circle surrounding each artificial seed cache, or "seed spot"; 3,804 seed spots were planted. The number of seeds planted per cache was not strictly controlled but was estimated to vary from 6 to 10. Most of the data (8 out of 10 sites) were obtained from experiments established for other purposes (e.g., tests of rodent repellents or shading). Seedling establishment improved progressively as the number of emerging seedlings increased from 1 through 8. Thus, clustered seedlings may have some intrinsic survival advantage over single seedlings. Holmgren and Basile suggested that the seed caches create a situation in which natural selection acts on the group of seedlings, and the surviving seedling is inherently better than those that died.[17] The expression of any inherent seedling survival characteristics would be influenced by so many environmental interactions, however, that it is doubtful that there would be any selective action. And if there were, it would be quickly lost in an obligate outcrossing population. The answer may simply be that multiple seedlings exceed the foraging capacity of a single vertebrate predator. At the time the study was conducted, seedling mortality resulting from excessive seedbed temperatures was a popular area of study in forestry. In that regard Ferguson and Basile suggested that multiple seedlings might increase the group's survival by shading one another and lowering the seedbed temperature. In reporting the results of a later study, however, Ferguson, downplayed the importance of excessive soil temperatures in seedling survival.[18] Ferguson and Basile's initial research was conducted on soils derived from decomposing granite—which tend to produce sand-textured surface soils with low moisture-holding capacity—that received about 15 inches of precipitation per year. Under natural conditions these sites would probably support mountain big sagebrush–antelope bitterbrush or ponderosa pine/mountain big sagebrush–antelope bitterbrush; certainly they do not fall on the arid end of antelope bitterbrush site potential.

In a subsequent study Ferguson measured the growth of single and multiple established antelope bitterbrush plants and found essentially no differences in the heights and diameters of individual versus multiple established antelope bitterbrush plants (no statistical comparison).[19] Fifty percent of seed spots (caches) having only one seedling alive at the end of the first year still contained a living seedling at the end of three years, while 75 percent of seed spots with multiple seedlings at the end of one year still contained living plants at the end of three years.

During the 1950s there was a great deal of interest in artificial seeding of antelope bitterbrush, and tremendous interest in seeding sagebrush ranges in general, for that matter. During the mid-1940s range managers determined that crested wheatgrass could be successfully seeded even on the arid sagebrush ranges of Nevada.[20] In order to convert degraded big sagebrush stands to perennial grass, it was necessary first to remove the woody sagebrush. Most of the stands did not have enough herbaceous understory to carry fires, so the brush had to be removed by other methods and a seedbed prepared mechanically. For this purpose an interagency equipment development committee developed the rangeland plow, using engineering expertise provided by the USDA, Forest Service. A succession of mechanical tools to aid in restoration of rangelands followed. The rangeland drill, perhaps the highest technological achievement of this effort, is still the basic implement for seeding rangelands.

Richard Hubbard was a strong proponent of using the new mechanical tools to prepare sites for seeding antelope bitterbrush.[21] In successful seedings in Modoc County, Hubbard first cleared away all the woody vegetation and then plowed and harrowed the site in the fall. The seedbed was allowed to settle until spring, when thiourea-pretreated seeds were planted. The paper that describes the project does not specify the method used to remove the woody vegetation, but it is noted as the most expensive portion of the operation. Most likely a combination of tractor-mounted bulldozer and handwork was involved. Hubbard put great emphasis on the importance of seedbed preparation in obtaining successful stands of antelope bitterbrush, perhaps because he had put considerable effort into determining the proper seeding depth.[22]

Hubbard arrived at two solid conclusions from this study. First, practically no seedling establishment occurred when seeds were simply broadcast on the soil surface. If they are to survive in semiarid to arid environments, seeds lying on the surface of the seedbed must take up moisture from the supplying substrate, with which they have only limited contact, faster than they lose moisture to the relatively dry atmosphere.[23] Furthermore, the seeds of antelope bitterbrush are considered a delicacy by several species of rodents, so exposure on the soil surface is a double hazard. Despite Hubbard's clear-cut finding, many hundreds of thousands (perhaps millions) of dollars worth of antelope bitterbrush seed has been broadcast and lost. Hubbard's second conclusion was that a planting depth of about 2 inches was the maximum depth for antelope bitterbrush seedling emergence. An occasional seedling would come up from 2½ inches, but clearly even 2 inches was too deep for optimum emergence. For that, seeds had to be placed about an inch deep in the seedbed. When continuous adequate moisture was present during the germination period, a satisfactory percentage of seeds buried less than an inch deep emerged, but there was a definite danger of the germination process being interrupted by moisture stress, especially in a coarser-textured seedbed with limited moisture-holding capacity. Basile and Holmgren reached similar conclusions in their seed spot trials in Idaho.[24] If the planting depth was too shallow, the seeds tended to be popped out on the soil surface at germination time, a circumstance the authors attributed to frost heaving. It seems unlikely, however, that such a level of frost action would occur during the germination period.

August Hormay reported in 1943 that rodents cache antelope bitterbrush seeds between ¼ and 1½ inches deep in the surface soil.[25] Does this reflect depth trials tested by generations of natural selection?

Antelope bitterbrush seeding techniques have not always followed standard methods. One seeding procedure formerly in wide use was the seed dribbler, a small seed box and metering system that was mounted on the top of one track of a track-laying tractor. Seed dribblers were frequently utilized on tractors equipped with bulldozer blades that were used to dress and construct water breaks on fire lines. As the tractor tracks moved, seeds fell on the pads and were pressed into the soil surface.

Mule deer habitat burned in wildfires is often reseeded by aerially broadcasting antelope bitterbrush seeds on top of the snow cover. In theory, the snow provides cover for the seeds for germination and seedling establishment. Wherever this technique has been tested, however, the end result has almost always been failure to obtain an established stand. There is some sense in seeding on snow. Being black and relatively dense, antelope bitterbrush seeds do selectively melt into the snow rather than skate along the surface. Germination occurs when the snow melts. But a seed exposed on the surface of the seedbed under snowmelt conditions is in an extremely harsh environment for germination. During the day it loses moisture to the atmosphere and at night it is exposed to freezing temperatures. We conducted extensive tests of antelope bitterbrush seeding on snow at Granite Peak and on the Dog Skin Mountains north of Reno. Although we tried broadcasting at different times (before, between, and after snowfalls) as well as different seed coatings and plant species, we were never successful in establishing anything except an occasional random seedling. Snowfall amounts and frequency made no difference. For any success, continuous warm (above freezing), moist conditions on the surface of the seedbed would have to obtain for the germination period after the snowmelt. It could happen, but such conditions are extremely rare in the environments to which antelope bitterbrush is adapted. Those who use the aerial broadcast method of seeding antelope bitterbrush on snow either believe in fairy tales or are not sufficiently well organized to get the seed into the ash immediately after the fire. Limited spot seeding by hand in the fall has a much greater chance of success than aerial broadcast seeding on the snow.

The purpose of site preparation is to reduce competition from existing vegetation and to prepare a seedbed in which antelope bitterbrush seeds can be precisely placed about 1 inch deep in the soil. This is easy to do in cultivated fields, but on wildland such precise placement is very difficult. Rocks, irregular microtopography, and woody debris all hamper seed placement. The rangeland drill was designed for use in such seedbeds, but it does not allow for precise seed placement or reliable seed coverage.[26] The maximum depth of each drilled opening can be controlled by placing depth bands on the disk openers. In very soft seedbeds, however, the seeds will be placed too deep even if depth bands are used. Planting seeds too deep with a rangeland drill is probably the most common reason for stand failures of numerous species, not just antelope bitterbrush.

The most common method of metering a drill's rate of seed distribution is a fluted shaft. The flutes on the turning shaft in the bottom of the drill box capture seeds as they turn and then carry them to a slotted, adjustable opening through which they drop into the seed tube to be conveyed to the opener. Antelope bitterbrush seeds are relatively dense, smooth, and slick. They do not pack or bridge over the fluted shaft as fluffy-seeded grasses do. The problem with antelope bitterbrush is that the rate of seeding is so low that it is impossible to calibrate. For example, crested wheatgrass is commonly seeded at the rate of 2 seeds per inch of row, which calculates to about 7 pounds per acre. Hubbard et al. suggested that an acceptable artificially seeded antelope bitterbrush stand should contain 1,500 plants per acre.[27] With a 12-inch spacing between the openers on a rangeland drill, that would be 29 feet between plants in a given drill row. Even in an excellent stand with 3,000 plants per acre there would be 14 feet between plants in a drill row. It is very hard to calibrate a drill to meter such low rates. In practice, many of the old rangeland drills *leak* more antelope bitterbrush seeds per acre than the desired seeding rate. Eamor Nord and Bert Knowles mixed rice hulls with the antelope bitterbrush seed at a ratio of eight parts antelope bitterbrush seed to three parts rice hulls (it is not clear if this was by weight or by volume).[28] Hubbard et al. suggested the same procedure but also failed to make it clear if the mixture should be by weight or by volume. Nord and Knowles tried for a seeding rate of 6–7 pounds per acre of antelope bitterbrush, and Hubbard et al. suggested 6 pounds. Using the rice hulls to dilute the antelope bitterbrush worked best if the mechanical agitation within the seed box was disconnected.

We have successfully substituted horticultural vermiculite for rice hulls. It is widely available at nurseries and plant supply stores and it can have only beneficial effects on the seedbed. Since the 1950s, when much of the antelope bitterbrush artificial seeding research was conducted, a host of native forb plant material has become available. Mixtures can be developed without including a competitive perennial grass, and the mixture itself used to help dilute the antelope bitterbrush seeding rate.

As we stated above, Nord and Hubbard both suggested a seeding rate of about 6 pounds per acre. There are about 15,400 antelope bitterbrush seeds per pound. That means they were seeding about 2 seeds per foot of drill row and considered 1,500 plants per acre a barely acceptable stand. Those figures give a 2 percent seedling establishment rate per seed planted. Even 3,000 antelope bitterbrush plants per acre in a good stand is less than a 4 percent success ratio. His observations of the first artificial seeding of antelope bitterbrush in the early 1940s led August Hormay to suggest that less than 5 percent of the seeds planted became established plants.[29] Perhaps Hubbard and Nord were referring to 6 pounds per acre of the rice hull–antelope bitterbrush mixture, and the mixing was done on a weight basis. This would give about an 8.6 percent success ratio. With antelope bitterbrush seed (cultivar Lassen) selling for about $20 per pound, that becomes an expensive, high-risk operation.

Describing the plant communities in which they worked was not the strong point of the Pacific Southwest Forest and Range Experiment Station scientists studying antelope bitterbrush in northeastern California, and that is unfortunate because it deprives later researchers of much useful information. Fortunately, other researchers have been more meticulous. Hubbard, for example, characterized his study site in Modoc County as transitional between ponderosa pine woodlands and the sagebrush/juniper complex.[30] The understory, which apparently was going to provide the competition in his studies of bitterbrush seedling survival, he described as perennial grasses and annual and perennial forbs. No mention was made of annual grasses or cheatgrass specifically. Hubbard used three treatments: (1) seeding antelope bitterbrush in the undisturbed native vegetation; (2) plowing, harrowing, and dragging with a rail before seeding to antelope bitterbrush alone; and (3) the same seedbed preparation, but seeding a mixture of antelope bitterbrush and crested wheatgrass. In the prepared seedbed plots only, a second set of treatments was superimposed: (1) weed free for three growing seasons, (2) weed free for one growing season, and (3) no additional weed control. Apparently, the plots were weeded by hand. Hubbard found only negligible weed competition in the prepared seedbeds the first growing season. The crested wheatgrass seedlings provided little competition the first year, but the second and third growing seasons they offered as much competition as the native vegetation.

Hubbard conducted this research in 1953 and published it in a research note in 1956. The basic results, as interpreted by game range managers, tremendously influenced mule deer range restoration during the second half of the twentieth century (Table 6.1). Our interpretation of this study and its subsequent influence is not meant to be critical of the author. At the time it was conducted, the research was appropriate and original. With the advantage of more than a half-century of hindsight, however, it becomes obvious that observations construed out of context can have major impacts on vast environments. Hubbard's observations were probably valid. Burgess L. Kay told us that he visited the site several decades after the seeding and could still pick out the an-

TABLE 6.1.
Mortality (%) through Three Growing Seasons of Bitterbrush Seedlings under Different Levels of Competition

Level of Competition	July 13, 1953	At End of First Growing Season	At End of Second Growing Season	At End of Third Growing Season
Negligible	17.1	21.0	22.4	22.4
Light:				
Weeded during 1st growing season	17.1	21.0	21.0	21.0
No weeding	17.1	20.5	20.8	20.8
Heavy:				
Crested wheatgrass	20.5	31.2	46.5	60.0
Native vegetation	17.8	45.0	55.0	56.6

Table reproduced as presented in Richard Hubbard, "The Effects of Plant Competition upon the Growth and Survival of Bitterbrush Seedlings" (Res. Note 109, USDA, Forest Serv., Berkeley, Calif., 1956).

telope bitterbrush plants in the plots that had been weeded for three growing seasons.

As was customary at the time, Hubbard did not statistically analyze his data, but some of the differences are probably significant. The mechanical tillage produced a clean seedbed, and weeding the first season was of no benefit that season, nor did it produce any carryover benefit for the next two seasons. Seeding antelope bitterbrush in the native vegetation reduced seedling establishment and continued to have an increasingly negative effect for the next two years. After the first growing season, crested wheatgrass competed with the antelope bitterbrush seedlings, influencing both seedling survival and antelope bitterbrush seedling height. Hubbard excavated root systems of seedlings from the different levels of competition and found very marked differences in root development (Fig. 6.7).

Hubbard's conclusion—that antelope bitterbrush should not be seeded along with other perennial grasses—is reasonable in the light of his data. He pointed out that there are valid reasons to plant perennial grasses for livestock forage

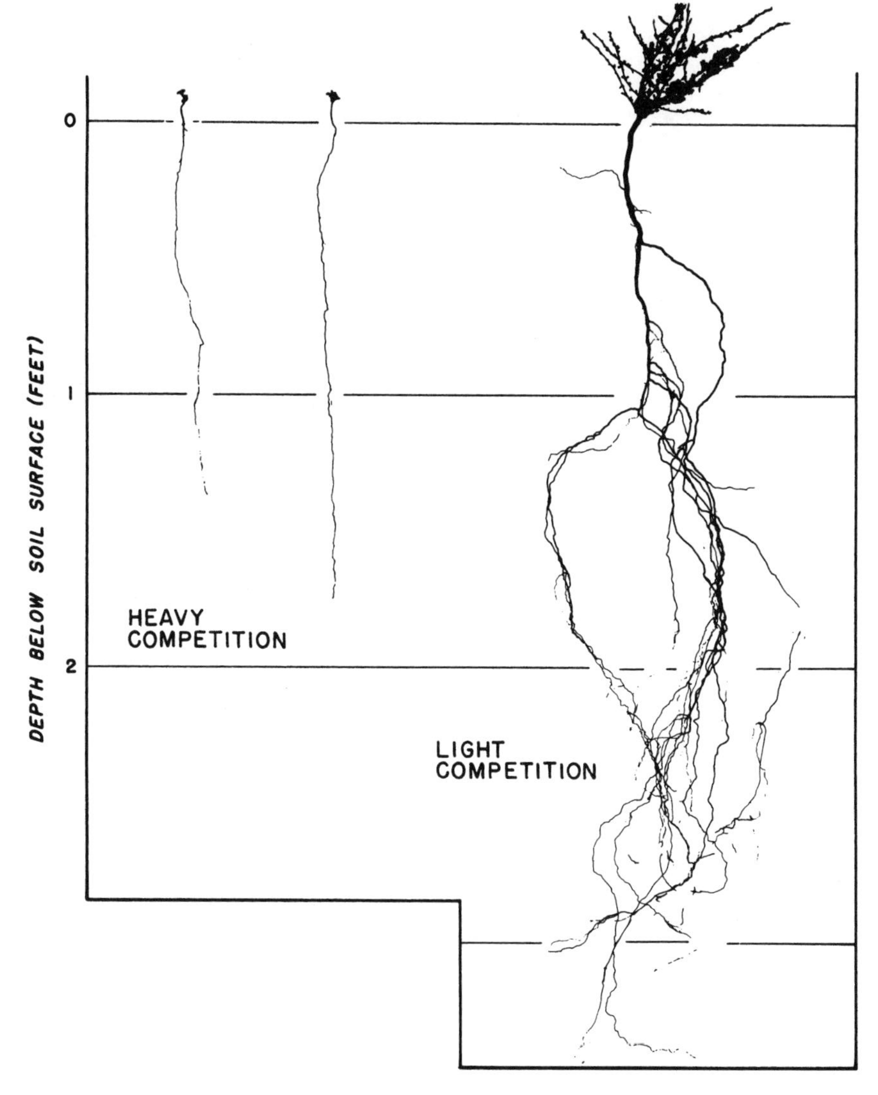

Fig. 6.7. Root development of antelope bitterbrush seedlings grown under two levels of plant competition. Adapted from Hubbard, "Effects of Plant Competition."

Fig. 6.8. Antelope bitterbrush seeded nine years previously in two strips, both plowed and harrowed before planting. The strip on the left was seeded with antelope bitterbrush alone and the one on the right was seeded to a mixture of bitterbrush and crested wheatgrass. From R. L. Hubbard, "A Guide to Bitterbrush Seeding in California" (Res. Note 34, USDA, Forest Serv., Berkeley, Calif., 1964).

and fire suppression. And he acknowledged that on the Paisley Ranger District of the Fremont National Forest in Oregon, antelope bitterbrush and crested wheatgrass had been successfully seeded together, although in alternate rows.[31] Perhaps most important, Hubbard noted that antelope bitterbrush could grow on many other types of sites than the one he studied. In 1964, Hubbard published an excellent photograph showing the difference between antelope bitterbrush seeded alone and seeded in a mixture with crested wheatgrass nine years before (Fig. 6.8). This is convincing evidence that competition is a major factor in antelope bitterbrush survival.[32] For some unknown reason, however, this is also a relatively site-specific phenomenon. We have 25-year-old plots of antelope bitterbrush and cliffrose at Granite Peak north of Reno, where crested wheatgrass and the shrubs were seeded together. Granite Peak, we should note, is a mountain big sagebrush / Thurber's needlegrass site that is much drier than Hubbard's Modoc County site. A fair summary of the available data would be that antelope bitterbrush and competitive perennial grasses should not be seeded in the same row, but the actual mechanism of competition and the site specificity of competition are not understood.

Over the years, the results of Hubbard's study came to be interpreted as meaning that antelope bitterbrush, and only antelope bitterbrush, should be planted in restoration projects. Game management agencies threatened to withhold funding or seed for wildfire restoration if any of the terrible crested wheatgrass was included in the seed mix. To understand what fostered this interpretation, it is necessary to remember the context of the time when this research was published. The post–World War II period was the golden age of range improvement on sagebrush rangelands. Millions of acres of degraded sagebrush were being plowed and seeded to crested wheatgrass for cattle forage. Game management was still a relatively young professional field in public land management agencies compared with forestry and range management. Many game managers believed that the crashes in mule deer numbers that were occurring in areas like the Devil's Garden in northeastern California were due to range improvement practices that converted shrub-dominated communities to crested wheatgrass. Crested wheatgrass thus became the symbol of antiwildlife management. One Great Basin wildlife manager told us that he became sick to his stomach when he drove by Hallelujah Junction on Highway 395 north of Reno because of the stands of crested wheatgrass. These stands were established in 1974 by crews under the direction of Richard Holland of the Bureau of Land Management during the restoration seeding of a 40,000-acre wildfire. Actually, the seeding is a mixture of several perennial grasses, two perennial legumes, and fourwing saltbush. Most of the seeded area did not support antelope bitterbrush before the wildfire. The portion of the burn in Sugar Loaf Basin was degraded antelope bitterbrush that was considered critical mule deer range. The basin is exactly on the California-Nevada state line. At the insistence of both state game management agencies the basin was seeded with unthreshed antelope bitterbrush seed only. The seeding was a failure and the cheatgrass-dominated basin has subsequently burned twice. Over the ensuing years, new generations of wildlife managers have arrived and departed, but the fear of seeding grass remains. It is doubtful that the current generation even remembers Hubbard's work.[33]

Land managers initially ignored Hubbard's discovery that native herbaceous perennial grasses and forbs had a negative influence on the establishment of antelope bitterbrush. In the almost half-century since then, the ecological condition of many upland areas in northeastern California has dramatically improved. It is now very apparent, as Hubbard experimentally demonstrated, that established herbaceous perennials, especially perennial grasses, close plant communities to the establishment of antelope bitterbrush seedlings. At the time Hubbard was working, the increase in native perennial grasses certainly was not a big worry. In the years since, however, game managers have been instrumental in decreasing livestock numbers on publicly owned ranges and enforcing the application of grazing management to these ranges. The results of this effort, especially at higher elevations above the zone of complete cheatgrass dominance, has been a slow but steady increase in the dominance of perennial grasses. Hubbard's observation that established perennial grasses have a negative influence on antelope bitterbrush establishment has been proven over a vast range of plant communities. "The Status of Antelope Bitterbrush in the Cassia Mountain Area of Southern Idaho," by range conservationist T. A. Phillips, does an excellent job of putting across this important point.[34] The paper describes an exceptionally well-designed and -documented study of antelope bitterbrush resources over time in a local area. There are a number of excellent figures comparing livestock numbers, mule deer numbers, and antelope bitterbrush establishment over time. As originally published, this work received limited regional distribution. It should be reprinted for wider distribution.

By the late 1950s, the excellent stands of antelope bitterbrush that Hubbard had established in Modoc County had begun to die. These stands had an enormous stocking rate—about 3,200 plants per acre. In his monograph on antelope bitterbrush Nord gave the average stocking rate for California as 780 antelope bitterbrush plants per acre and the maximum for a natural stand as 1,420. For reasons that are not completely clear to us, in their 1962 paper Hubbard et al. recommended *slightly below* 2,200 antelope bitterbrush plants per acre as the minimum acceptable stand density for artificially seeded stands. They also concluded that crested wheatgrass had to be at least 2 feet from antelope bitterbrush plants to prevent competition. The authors did note, citing the work of Howard R. Leach, that mule deer eat grass at certain times of the year and suggested that perhaps there should be something in restored mule deer range in addition to antelope bitterbrush.[35]

In the summer of 1942, a wildfire burned approximately 1,200 acres of rangeland in the Willow and Case Creek drainages of the Mountain Home District of Boise National Forest in Idaho. Prior to burning, the site supported a mixed stand of mountain big sagebrush and antelope bitterbrush with a bunchgrass understory. About 45 acres of the burn was seeded to a tall upright form of antelope bitterbrush and a grass-forb mixture. The antelope bitterbrush was planted separately by blocking the outside portion of one drill box. The site was protected from grazing for two years and then subsequently grazed by cattle. When the first count of antelope bitterbrush seedlings was made in 1949, nine years after the seeding, one plant had become established for every 7.6 feet of drill row. Steve Monson and Nancy Shaw resampled the site in 1981 but did not calculate a density of antelope bitterbrush plants per acre because the stand consisted of a single row of bitterbrush separated by 20-foot strips of herbaceous vegetation.[36] If it had been a solid stand, with 1-foot drill row spacing, the antelope bitterbrush population would have been 5,700 plants per acre, a very dense stand. More than 99 percent of the plants recorded in 1949 were still alive in 1981. Light to moderate livestock grazing had markedly reduced the antelope bitterbrush population relative to that in the protected areas. The site received little mule deer use. The protected shrubs within the grazing exclosure were about 15 percent larger than their grazed counterparts outside the exclosure. There were 1.3 times as many natural antelope bitterbrush seedlings established inside the grazing exclosure as outside (205 versus 149). In comparing the occurrence of antelope bitterbrush plants and understory grass, Monsen and Shaw determined that the native grasses Idaho fescue (*Festuca idahoensis*) and Sandberg bluegrass (*Poa secunda*) were negatively related to antelope bitterbrush cover, and squirreltail (*Elymus elymoides*) was positively related.

Hubbard never attached much importance to cheatgrass. Apparently it was not a problem during the 1950s at the higher elevation experimental sites in Modoc County, although some of the Lassen County study sites such as Red Rock and Doyle must have been infested with it by then. Hormay mentioned cheatgrass in 1943. In a general paper published in 1948, Hurd noted that cheatgrass inhibited the establishment of antelope bitterbrush.[37] The first paper that was specifically written about antelope bitterbrush–cheatgrass competition was by Ralph Holmgren based on research conducted in Idaho.[38]

Holmgren was aware of the brilliant research conducted by R. L. Piemeisel on secondary succession in degraded sagebrush communities and eventual cheatgrass dominance.[39] Piemeisel worked on the Snake River Plains and described the succession from bare ground through Russian thistle and annual mustards to cheatgrass dominance. Holmgren was also aware of Joseph H. Robertson's classic paper on cheatgrass dominance and knew that once cheatgrass was present in a plant community, other perennial seedlings would not become established.[40] Robertson's research applied to the establishment of exotic perennial grass seedlings, but Holmgren made it clear that the extremely competitive cheatgrass effectively closed communities to the establishment of antelope bitterbrush seedlings as well.

Holmgren's answer to cheatgrass competition was to physically remove the surface soil from around a spot where antelope bitterbrush seeds were buried in a cache or seed spot. The process is called *scalping*. Because cheatgrass is an annual, the scalping process destroys germinated seedlings. Cheatgrass builds up huge seed banks, but the seeds are located at or very near the soil surface and in the annual grass litter.[41] In the fall, Holmgren scalped the plots to a depth of 2 inches in diameters of 1, 1½, 2½, and 4 feet. The percentage of seedling survival and the height of the antelope bitterbrush seedlings increased with increasing size of the scalped area. Holmgren later estimated that 12 man-hours were required per acre for scalping and hand seeding.[42] In the 1950s, when Holmgren was first working on antelope bitterbrush–cheatgrass competition, he did not think that cheatgrass developed seedbanks.[43] This was the general

Fig. 6.9. Hansen scalper–browse seeder. From Giunta et al., *Antelope Bitterbrush.*

consensus at the time because freshly harvested seeds of cheatgrass would germinate with no dormancy. During the 1960s, however, we demonstrated that cheatgrass seeds that do not find safe sites for germination can *acquire* dormancy in field seedbeds.[44]

Later, the scalping process was mechanized on sites where topography and rocks permitted the use of a tractor (Fig. 6.9). The large flat-lister type shovel used for the scalping eventually evolved into the Hansen browse seeder. In fact, the widespread application of scalping for antelope bitterbrush seeding fostered the development of a range of specialized equipment to make the process easier, including a scalping hoe manufactured from a square-nosed shovel and the Schussler single-handed spot seeder, a modification of a corn planter developed by Howard Schussler (Fig. 6.10).[45]

On a mule deer winter range near Manti, Utah, B. C. Giunta et al. experimented with a mechanical scalper-seeder in trials that included antelope bitter-

Fig. 6.10. Schussler antelope bitterbrush spot seeder. From Giunta et al., *Antelope Bitterbrush.*

brush and cliffrose,[46] comparing 4-, 8-, 16-, and 24-inch-wide scalps. Antelope bitterbrush and Nevada ephedra (*Ephedra nevadensis*) showed a positive response to increasing width of scalping. Cheatgrass reestablished in the scalps, however, and suppressed the shrub seedlings that did establish.

Hubbard believed that once antelope bitterbrush seedlings reached a certain size, apparently at several years of age, they could suppress herbaceous competition rather than being suppressed by perennial herbaceous species. Perhaps he reached this conclusion by observing the many old-growth antelope bitterbrush communities that still existed at midcentury where that rule apparently held true. These stands had established in the last decades of the nineteenth century or very early in the twentieth when sagebrush-bunchgrass ranges were often severely overgrazed. Most important, the stands had become established before cheatgrass gained widespread dominance. The paper by Giunta et al. includes a photograph of the established browse species. For antelope bitterbrush, an arrow had to be superimposed to show where the shrubs were hidden in the returning cheatgrass. Not only were the shrub seedlings suppressed by cheatgrass competition, they were each year in danger of being killed by a wildfire. Shrubs do not completely suppress cheatgrass; only perennial grasses can do that.

Cheatgrass is one of the devils that dog projects to artificially restore antelope bitterbrush. Another is bringing young stands of planted antelope bitterbrush to a stage at which they supply adequate forage for deer under persistent heavy browsing. J. Edward Dealy examined this issue in his studies of the Silver Lake mule deer ranges in Oregon.

A couple of years before Dealy's work was pubished, Robert Ferguson reported on the survival and growth of planted bitterbrush in Idaho over 9- and 15-year periods.[47] In one area he found 20 percent greater shrub survival without browsing; in another he found no difference in survival between protected and browsed plants growing under similar conditions. Smaller plants that were competing with cheatgrass were more likely to be damaged by browsing than vigorous plants with less competition, however. Ferguson suggested that removal of one-third to one-half of the current annual growth by browsing mule deer would kill antelope bitterbrush plants that otherwise would have survived under protection.

Dealy set up four comparisons in the Silver Lake study site: (1) caged (to prevent browsing) with competing herbaceous vegetation, (2) caged with competing vegetation removed by hand, (3) uncaged with competing vegetation, and (4) uncaged with no competing vegetation.[48] After two years, the plants protected from competition showed a positive response in shrub height averaging 0.7 feet. Those protected from browsing showed a much smaller increase in diameter. Protection from browsing was thus more important in increasing crown cover than removal of competition.

During the 1960s and 1970s there was a lot of research on the use of herbicides to reduce competition on rangelands and improve the success of seedings. Some of the most successful research on the use of herbicides to control cheatgrass was conducted by Raymond Evans and Dick Eckert, who developed two approaches to suppression. One method was to use a contact herbicide that was deactivated as soon as it contacted the soil. This treatment killed only cheatgrass plants that had germinated; therefore it had to be applied in early spring and followed immediately by seeding. The second method involved applying a soil-active herbicide that would keep the site free of cheatgrass for a year, a herbicidal fallow. In practice the herbicide would be applied in the fall and the site seeded the next year. This was not a simple operation. The herbicide had to have a broad enough spectrum of weed control to kill cheatgrass and the associated broadleaf species, but at the same time, all biological active herbicide residue had to be dissipated by the end of the fallow period when the forage species were seeded. Evans and Eckert tested numerous compounds before they decided on Atrazine for the herbicidal fallow and Paraquat for the contact herbicide.[49]

By the time these weed control methodologies had made it through the tortuous federal approval process, the golden age of range improvement was over. It was replaced by an emphasis on grazing management to solve the inherent problems of sagebrush-bunchgrass ranges in a political atmosphere of distrust of pesticide use in general. The registration for use of both products has been dropped because use has not justified their renewal. The principles Evans and Eckert developed and tested remain valid. Any use of modern herbicides to control cheatgrass to aid in the establishment of antelope bitterbrush must fall within the same framework of treatments unless a truly physiologically active herbicide is developed.

One of the very interesting aspects of the herbicidal fallow technique for controlling cheatgrass was the discovery that a year of fallow did not destroy the cheatgrass seed bank. However, removing the litter cover on the surface of the seedbed created an environment in which the cheatgrass had difficulty germinating because it was losing moisture to the atmosphere faster than it could take up moisture from the substrate.[50] The herbicidal fallow, that is, induced physical changes in the quality of the seedbed that limited reestablishment of cheatgrass. The seed of perennial grasses was placed, through the drilling process, in an environment favorable for germination. This is in essence the same general process as mechanically scalping spots and seeding antelope bitterbrush in seed spots.

Herbicidal fallows created with Atrazine proved to be excellent seedbeds for the antelope bitterbrush seeds cached by rodents.[51] And seedlings of antelope bitterbrush could be transplanted into the Atrazine fallow in the spring following the application of the herbicide. The herbicide was active enough to control annual grasses, but it did not harm the antelope bitterbrush seedlings. This is not to say that the Atrazine showed physiological selectivity; the roots of the transplants were below the surface soil where the highly insoluble Atrazine was located.

Of the major species that are artificially seeded on sagebrush and related rangeland environments, antelope bitterbrush has the lowest probability of success. The requirements that must be met to establish seedlings of this species are daunting. Seedling growth subsequent to initial establishment also has demanding requirements. The current challenge to mule deer habitat managers is to find a way to recruit significant numbers of antelope bitterbrush seedlings into communities closed by native perennial or exotic annual grasses.

Chapter Seven

Granivore Relations

Granivores are animals that depend on seeds as a major component of their diet. Ground squirrels, chipmunks, rats, mice, various birds, and harvester ants are all examples of granivores. Granivory, the consumption of seeds by animals, can limit the establishment of plant species through seed predation, but it can also be essential in seed dispersal and seedling recruitment.[1]

Under the right conditions, *Purshia* species can produce large numbers of viable seeds with the potential for copious seedling recruitment. Once seeds disperse from the maternal plant they must find a safe site for germination within the seedbed, a site where the physical and biological parameters of the environment match the inherent potential of the seeds to germinate. Remember that the *Purshia* achenes require a period of cool-moist pretreatment before their dormancy is broken and germination can occur. Because the achenes must be fully imbibed for dormancy to be broken and prechilling is most effective between 36 and 40°F, dormancy is most effectively broken when the seeds are buried in the surface soil. Burial dampens temperature extremes and enhances moisture relations for the seed. Predation by granivores directly affects the number of seeds (achenes) that reach safe sites for germination and the physical status of those seeds.

The earliest researchers on the natural and artificial regeneration of antelope bitterbrush, August Hormay and Eamor Nord, found granivorous rodents to be a very significant factor in seed dispersal and seedling establishment. Despite that discovery, however, a tremendous amount of research has been devoted to overcoming perceived rodent predation on antelope bitterbrush seeds.

Rodents are not the only significant predators of *Purshia* seeds. Insects and birds also play a role in bitterbrush and cliffrose seed ecology. Perhaps the most active and efficient group of seed predators in North America are the harvester ants of the genus *Pogonomyrmex*.[2] Seed-harvesting ants are important components of arid and semiarid shrub-steppe and desert ecosystems. They have been known to completely remove the vegetation surrounding the cone of their nests to a radius of up to 6 feet.[3] The distribution of *Pogonomyrmex* blankets the arid regions of Mexico and the western United States. The worker ants collect seeds for food either before they are dispersed from the plant, by cutting the seeds away with their mandibles, or from the surface of the seedbeds beneath the can-

Fig. 7.1. Typical harvester ant mound surrounded by area cleared of all vegetation. The ants' seed-harvesting and seed-processing activities often create unique plant communities on portions of the outer mound.

opy of the maternal plants. The seeds or fruits are often processed before storage. From certain seeds, for example, the ants remove a special appendage, the eliasome, which contains an oil droplet that serves as a food source for the ants. After the eliasomes are removed, the seeds are discarded on the soil surface in dumps surrounding the mounds.[4] Ants may also gather seeds and store them on the soil surface on and around the mound, perhaps drying them before they are transported to the subterranean storage chambers. As harvester ants change the physical characteristics of the soils surrounding their mounds, this in turn creates specialized seedbed characteristics. The seed-caching activities of the ants and the characteristics of the mound soils attract granivorous rodents.[5] Whatever the reason, harvester ant mounds often have a specialized flora of monospecific microcommunities that reflect the ants' interactions with seeds and seedbeds (Fig. 7.1).

The mass of antelope bitterbrush seeds (about 0.0227 grams per seed) is near the limit that harvester ants can transport long distances, although it has been reported that the seeds are too heavy for the ants to carry. Raymond Evans et al.

reported predation of antelope bitterbrush seeds by harvester ants,[6] and we have observed ants harvesting the seeds and carrying them into their underground nests. We have also observed harvester ants carrying antelope bitterbrush seeds and dropping them before reaching the nest. Such abandoned seeds are dispersed without consumption, a boon to the plant. Virtually nothing has been documented regarding the influence of harvester ant predation on bitterbrush and cliffrose seeds, but anyone who has ever witnessed the incredible ability, strength, and determination of ant colonies will realize the importance these animals can have for a given environment.

The extent to which birds prey on *Purshia* seeds is not very well known either, although nutcrackers (*Nucifraga columbiana*), pinyon jays (*Gymnorhinus cyanocephala*), and shrikes (*Lanius* spp.) are known to harvest them.

Granivorous rodents are probably the most important actors in the seed and seedbed ecology of *Purshia* species for three reasons: (1) they remove the remnant style, reducing seed dormancy in antelope and desert bitterbrush; (2) they aid in seed dispersal; and (3) they cache seeds in scatter hoards in the surface soil, which in advantageous microenvironments allows for prechilling and subsequent germination.

Purshia seeds are cached by rodents belonging to such families as Sciuridae, chipmunks and ground squirrels (diurnal animals); Heteromyidae, kangaroo rats and pocket mice (nocturnal animals); Cricetidae, woodrats, and voles; and Muridae, deer mouse and pinon mouse (Fig. 7.2). Rodents in the family Heteromyidae have fur-lined external cheek pouches, an adaptation that allows them to harvest a substantial amount of seed (Fig. 7.3). Rodents harvest the seeds, cache them, and return to harvest more seeds. The number of seeds harvested at one time depends on the size of the animal, its physical attributes (cheek pouches) and behavior, and the size of the seeds being collected.

Granivorous rodents are known to be very active both in harvesting antelope bitterbrush seeds and in dispersing the seeds through their caching behavior.[7] Two types of caching have been described: *larder-hoard caching,* in which seeds are cached deep within the burrow; and *scatter-hoard caching,* in which seeds are buried in shallow, scattered depressions throughout the animal's home range (Fig. 7.4). Like those in harvester ant storage chambers, seeds in larder hoards are buried too deep for emergence. Edward Schneegas reported finding an antelope bitterbrush seed cache (larder hoard) in a hollow aspen (*Populus* spp.) that yielded 10 pounds of cleaned bitterbrush seed, or 170,000–200,000 individual antelope bitterbrush seeds.[8]

The relationship between granivores and bitterbrush is very interesting, especially when scatter-hoard caching by rodents is examined more closely. Vander Wall and Evans et al. both noted that antelope bitterbrush seeds benefit by being cached. Antelope bitterbrush seeds that are cached are better dispersed,

Fig. 7.2. Ord's kangaroo rat (*Dipodomys ordii*) is in the family Heteromyidae.

cached at favorable depths for germination, and are less likely to be preyed on by other granivores such as birds and harvester ants. Granivorous rodents protect their caches from robbery by other rodents of the same or different species. They often dig up their caches and move them to confuse other potential predators, but in the process some seeds are consumed. The ability of granivorous rodents to find buried seeds varies according to the species and the physical conditions (e.g., soil moisture content) of the seedbed. Scatter-hoard caches are often either forgotten or not fully eaten by the rodent cachers, and therefore can be very important in the recruitment of seedlings.

Steve Vander Wall conducted an extensive research experiment in Little Valley, Washoe County, Nevada, during 1988–1992 with rodents and antelope bitterbrush. Ninety-nine percent of the antelope bitterbrush seedlings in this Jeffrey pine–antelope bitterbrush community resulted from rodent caches.[9] The vast majority of caching in the area he studied was done by yellow pine chipmunks (*Tamius ameonus*), which cached the seeds at depths of less than

Fig. 7.3. Fur-lined external cheek pouches typical of members of the family Heteromyidae. The pouches are used to transport collected seeds to cache sites.

Fig. 7.4. Antelope bitterbrush seedlings emerging from a rodent-created scatter-hoard cache.

1 inch in the surface soil. Vander Wall reported that rodents harvested nearly all of his isotope-labeled seeds within 24 hours.

We conducted research on the harvest and consumption of antelope bitterbrush seeds by rodents at the Doyle Wildlife Management Area. The rodents most frequently encountered at this big sagebrush–antelope bitterbrush/cheatgrass site were Ord's kangaroo rats (*Dipodomys ordii*), deer mice (*Peromyscus maniculatus*), and Great Basin pocket mice (*Perognathus parvus*). These three species harvested 85 percent of the antelope bitterbrush seeds available to them within a 24-hour period and consumed 75 percent of those seeds within that time. It should be noted that these rodents were confined to portable enclosures, which may have been responsible for the high percentage of antelope bitterbrush seeds consumed.

Early researchers like Nord and Hormay also noted that the vast majority of antelope bitterbrush seedlings emerged from rodent caches. Hormay observed in the early 1950s that less than 5 percent of the total antelope bitterbrush seed crop produced seedlings, and Vander Wall, working 40 years later, found that only 2.5 percent of the total seed crop produced seedlings. These numbers, combined with granivore seed predation, may render seedling establishment insufficient to renew stands in many habitats.

Decadent antelope bitterbrush stands without seedling recruitment are known from several locations.[10] Researchers in southern Oregon found that the average antelope bitterbrush establishment was 0.7 plants per acre per year from 1925 to 1975. This area averaged 473 antelope bitterbrush plants per acre, and maintaining that density level of plants would require the establishment of 6.7 plants per acre annually.

Vander Wall suggested that antelope bitterbrush and granivorous rodents may have evolved a mutualistic association.[11] Antelope bitterbrush might once have depended on frugivorous birds and mammals to be dispersal agents, but scatter hoarding became a more efficient means of seed dispersal. It is in any case apparent that rodents play an important role in the handling and dispersal of antelope bitterbrush seeds.

The number of antelope bitterbrush seeds in a scatter-hoard cache is variable. Vander Wall found from 1 to 23 seeds, with an average of 10–12, per rodent cache in the woodland pine/antelope bitterbrush community he studied. He also found a few scatter-hoard caches containing 200–300 seeds. Merriam's kangaroo rats, which are common in big sagebrush–antelope bitterbrush communities, have been reported to cache an average of 33 antelope bitterbrush seeds per cache.[12] S. R. Morton et al. reported that most heteromyid rodents can collect more than their daily requirement of seeds in a single trip.[13] The external cheek pouches of heteromyids increase in capacity as their body mass increases.

The size of scatter-hoard caches is important in relation to seed production and seed dispersal. The fewer the antelope bitterbrush seeds produced in a given year, the lower the number of seeds for germination and emergence of seedlings the following spring. A certain percentage is going to be eaten by granivores, and this percentage, according to Vander Wall, is similar every year, regardless of seed production.

Depending on the year's seed production, the smaller the cache size, the better the distribution of the scatter hoards. Therefore, it would benefit the plants for rodents that cache antelope bitterbrush seeds in smaller numbers to be the dominant species on sites that have poor seed production. Unfortunately, the sites that are experiencing poor seed production are frequently critical mule deer winter ranges located at the lower elevations in sagebrush communities. Kangaroo rats, which make large caches, are more common in those environments than chipmunks, resulting in less seed dispersal in those critical habitats. The larger the cache size, the greater the potential for intraspecific competition for seeds. And since sagebrush-steppe environments do not experience season-long snow cover, these rodents forage on seeds in their scatter-hoard caches rather than eating seeds in their larder caches, which are unavailable for emergence anyway.

Because granivorous rodents do not harvest sagebrush seeds, which are very small, their preference for antelope bitterbrush seeds increases in sagebrush-bitterbrush communities (preference is often related to availability; therefore the "wasted" space taken up by sagebrush increases rodents' preference for antelope bitterbrush seed in that community). In ponderosa and Jeffrey pine woodlands, in contrast, pine seeds are a major food source for rodents, reducing the pressure on bitterbrush seeds.

Deer mice are common in sagebrush-steppe environments (Fig. 7.5). During our work at the Doyle Wildlife Management Area we have observed high deer mouse predation on artificially planted antelope bitterbrush seeds in both caches and drill rows. In one trial at Doyle, we used a rangeland drill to create drill rows in a relatively large area without dropping any seed. We then went back and seeded by hand short sections of the drill rows that had been previously selected at random. Our hypothesis was that once deer mice discover antelope bitterbrush seeds in a drill row, they proceed down the row and harvest everything in the row. To our surprise, the widely spaced, randomly located seeded sections were almost totally stripped by rodents. Deer mice are reported to be severe seed predators of several other revegetation species as well.

Rodents' preference for antelope bitterbrush seeds is not merely a measure of their harvesting and consumption rates; the nutritional value of the seeds is also important. Desert rodents depend on seeds for their various nutritional requirements (e.g., protein and fat) as well as for water. Kelrick and MacMahon and Kelrick et al. studied native seed nutrition for granivores in a shrub-steppe

Fig. 7.5. Deer mice (*Peromyscus maniculatus*) are a common granivorous species in sagebrush–antelope bitterbrush communities.

environment in Wyoming and found that *Purshia* seed was preferred.[14] Bitterbrush seeds had the highest percentage of soluble carbohydrate and water, and the lowest percentage of structural carbohydrates (which require more water for fecal excretion). The moisture content of a seed type can influence certain rodents,[15] but soluble carbohydrates may be one of the most important traits determining rodent seed preferences.

We have been tracking the nutritional status of antelope bitterbrush seeds during the winter months (October through early March) at the Doyle Wildlife Management Area and looking for a correlation with rodent preferences. We placed antelope bitterbrush seeds in nylon net bags and buried them in the surface soil, then recovered the bags monthly and compared them with control seeds stored dry at 40°F. Antelope bitterbrush seeds underwent physiological changes very rapidly after the seeds imbibed water in the cold seedbeds. The seedbed did not have to be saturated for the seeds to imbibe sufficient moisture for prechilling to break dormancy. Root emergence occurred as early as November. However, straightening and elongation of the hypocotyl arch (the phase of germination visible aboveground) did not occur until March. Concurrent with this early root emergence, nutritional analysis revealed an increase in carotene. Beta-carotene is an essential dietary compound because it is a metabolic precursor of vitamin A. Vitamin A deficiency deters growth and reproduction. Although adult mammals store vitamin A in the liver, newborns lack such stores, and juveniles therefore require at least moderate amounts of beta-carotene for survival.[16] The dietary intake of beta-carotene can thus have important fitness consequences in rodents. Vitamin A is also essential in the production of the visual pigment rhodopsin. Deficiency of this visual pigment can eventually lead to the loss of vision, but perhaps equally important for nocturnal desert rodents is the loss of the eyes' ability to accommodate to darkness, impairing night vision. Kangaroo rats, and probably other nocturnal rodents as well, rely on sensitive night vision to detect and escape from predators.[17] Desert rodents' dietary requirements for beta-carotene are not known, but laboratory rats require 4–6 milligrams per kilogram of body mass per day. If we consider this figure a guideline, a 20 gram deer mouse would require 0.08 milligrams per day of beta-carotene, and an 80 gram kangaroo rat would require 0.32 milligrams per day. The rodents would have to ingest 40 and 160 grams, respectively, of nonprechilled antelope bitterbrush seeds per day to meet these requirements, which greatly exceeds the normal daily intake. However, antelope bitterbrush seeds that have been naturally prechilled (i.e., placed into the soil) would satisfy the requirement with 0.15 and 0.58 grams of seeds, one-fourth the amount of seeds these rodents can consume per day. Prechilling is thus nearly as important for seed predators as it is for the seeds. In our study of rodent preferences, the Panamint kangaroo rat (*Dipodomys panamintinus*) ate more prechilled antelope bitterbrush seeds than control seeds, and the consumption of both types of seeds increased month by month. The fact that the consumption of control seeds also increased suggests that the increase in consumption of prechilled seeds is not the result of an increase in a certain nutritional component such as beta-carotene, but is instead driven by the aversion to some component of nonprechilled seeds such as fiber, which is higher than in prechilled seeds. This aversion is reduced through the prechilling process and resulting physiological changes in the seeds.

Jenkins and Ascanio reported that kangaroo rats refrain from using antelope bitterbrush seeds.[18] Kangaroo rats at our Doyle field site and in the laboratory consumed both prechilled and control antelope bitterbrush seeds, however, sometimes in large quantities (100 seeds in 24 hours). The moisture content of antelope bitterbrush seeds increases as they imbibe after they are cached in the soil. The cold nights of late autumn cause moisture condensation from the air on the surface soil particles even before the first biologically effective moisture event. The increase in moisture content of the seeds combined with the consumption of other seeds, insects, and succulent foliage that contain water may relieve the kangaroo rats' water stress (which is caused by the negative relation between protein and/or fat intake to metabolic water assimilation) and allow them to use antelope bitterbrush seeds as a food source.

Jenkins and Ascanio also reported that the condition of the kangaroo rats used in their study deteriorated when they were given a 100 percent antelope bitterbrush diet. This is not surprising, as rodents—and other animals as well—do not benefit from a 100 percent diet of any seed species, but rather require a variety.

Range managers began their fight against rodent seed predators in the 1950s by trapping rodents in areas they were preparing to seed. This was not a successful strategy, as rodents from neighboring areas invaded the habitat faster than they could be removed. Donald A. Spencer, a biologist with the U.S. Department of Interior, Fish and Wildlife Service, wrote *Rodents and Direct Seeding* during that period. Among other things he reported that as few as six deer mice per acre could largely nullify a program in which Douglas fir (*Pseudotsuga menziesii*) seed was broadcast at the rate of ¼ pound per acre.[19] Spencer also reported that trapping was ineffective. During the summer months, rodents could repopulate even large areas (i.e., 100 acres) in as little as 30 days. For example, a 5-acre tract in southern Arizona had an estimated resident population of only 15 gray-tailed antelope squirrels (*Citellus harrisi*), but 119 were trapped there over a 10-month period. On this same area there were an estimated 107 white-throated wood rats (*Neotoma albigula*), but regular biweekly trapping for 10 months removed 682 of these rodents.

The failure of poisoning and trapping to remove rodents led to the development of chemical compounds to prevent seed predation. The basic concept involved the use of a nonlethal seed treatment that would repel rodents. Tetramine (tetramethylene disulphotetramine) was the first chemical discovered for this purpose. Spencer reported that its use in the early 1950s decreased rodent seed predation in direct seeding efforts. He recommended that antelope bitterbrush be treated by immersing dry seeds in a 1 percent acetone solution of tetramine for one hour and then drying for at least one hour. Tetramine was not the ideal solution to the problem, however. All deer mice that ate 5 or more seeds died. Caged deer mice would consume an average of 17 tetramine-treated Douglas fir seeds when the seed was initially offered.

The toxicity continued after the seeds germinated. Seedlings were toxic enough to kill meadow mice (*Microtus* spp.), and the toxicity remained in the food chain, killing animals that scavenged on the dead meadow mice. Usually, the animals at the next level became ill from the tetramine transfer but did not die.[20]

Robert Casebeer, who was with the Idaho Fish and Game Department in the 1950s, also studied tetramine use in deterring rodent predation.[21] When he planted tetramine-treated antelope bitterbrush in artificial caches, the level of rodent predation was only 5 percent. He compared the emergence of tetramine-treated and control seeds planted inside cages to deter rodents to determine if the chemical had an influence on germination. There was no adverse effect; both the treated and untreated seeds exhibited 87 percent emergence. He then repeated the experiment without the protective cages. Only one cache that was treated with tetramine was disturbed, whereas all the untreated caches were disturbed. Antelope bitterbrush seedling emergence was 86 percent from the treated caches and less than 1 percent in the untreated ones.

Casebeer monitored the rodent population through live-trapping and tagging both before and after using the tetramine. The use of this compound did not eliminate the rodent population, because he captured similar numbers both before and after it was used. The number of tagged rodents he recaptured, though, was quite low, which could indicate that rodents from outlying areas were replacing rodents that died from natural causes or tetramine poisoning.

This experiment illustrates an unexplained aspect of apparent rodent predation on antelope bitterbrush seeds: the great variability over time at a given site. Perhaps the species composition and/or density of rodents suddenly changes, or maybe the rodents become familiar with seeding practices and change their behavior to take advantage of the seeded species.

A strategy of providing a more preferred seed—a decoy seed—along with the antelope bitterbrush seeds has recently been tried. R. L. Everett et al. studied deer mice seed preferences using commonly planted species (antelope bitterbrush, indigenous weed seed [cheatgrass, *Bromus tectorum*], and decoy seed wheat [*Triticum sativum*]). Antelope bitterbrush was the most preferred seed, comprising as much as 32 percent of the rodents' diet.[22] Singleleaf pinyon (*Pinus monophylla*) was the second most preferred seed, at 26 percent. The preference for antelope bitterbrush decreased when more seeds of singleleaf pinyon were available, and also when the seeds were coated with a rodent pesticide-repellent (alpha-naphthylthiourea).

Evans et al. also found that rodents preferred antelope bitterbrush seeds when these were offered in combination with other seeds.[23] Rodents did not disturb caches of 100 seeds of crested wheatgrass. When the crested wheatgrass seeds were mixed with finely ground antelope bitterbrush seeds before caching, however, the rodents completely excavated the seed boxes and chewed up the fiberglass screens that had been placed below the seeds to aid in their recovery. Apparently, they were attracted by the smell of the antelope bitterbrush seeds.

Vander Wall noted good antelope bitterbrush seedling recruitment and healthy stands at his study area in the eastern Sierras in pine woodland/antelope bitterbrush communities (30–70 antelope bitterbrush seedling clumps per 1,000 square feet). We have reported, along with others, the poor antelope bitterbrush seedling recruitment (finding 20 clumps in a full day's search would be extremely good) in older, more decadent stands in big sagebrush–antelope bitterbrush communities at the lower elevation sites along the trans-Sierra front. The lower elevation sites rarely have continuous snow cover, whereas the higher elevation sites frequently experience season-long snow cover. We see this as an important aspect in granivore–antelope bitterbrush relations because the lack of snow cover ensures that rodents will visit their seed caches frequently throughout the winter months. Scatter-hoard caches at higher elevations, in contrast, benefit from the limited rodent activity (i.e., hibernation) and excellent conditions for prechilling to break dormancy.

We previously mentioned that 99 percent of the antelope bitterbrush seedlings Vander Wall found at his pine woodland/antelope bitterbrush site resulted from rodent caches. West reported that 50–90 percent of the antelope bitterbrush seedlings at his pine woodland/antelope bitterbrush site in central Oregon resulted from rodent caches. Ferguson found that a single seedling had less chance of becoming established than multiple seedlings (two or more seedlings per clump). Vander Wall suggested that large clumps of antelope bitterbrush seedlings are significantly more likely to be represented by one or more living seedlings by the end of their first growing season than are small clumps. He also reported that a significant number of clumped seedlings survived, whereas single seedlings experienced a high mortality rate. The reasons for this significant difference in seedling survivability are not well understood, but multiple seedlings may benefit from the increased shading (decreasing moisture stress). Multiple seedlings tend to emerge from caches placed in the open interspaces between shrubs (decreasing competition for resources such as moisture) rather than under the canopy of an existing shrub. The seedlings' competition among themselves for resources would seem to negate any moisture relation advantage gained from mutual shading. Apparently, no one has ever compared the rate of emergence of single versus multiple seedlings. The seed caches may have a better environment for prechilling than a single seed on the seedbed surface, so that emergence, or at least root growth, occurs earlier in caches.

Rodents placed 49 percent of their antelope bitterbrush seed caches under bitterbrush shrubs, 34 percent at the margins of the bitterbrush shrubs (within 4 inches of the canopy edge), and 17 percent in the open. This would suggest that the majority of the caches had to compete with the existing shrubs for moisture and nutrients. Vander Wall further reported that the majority of the seedlings under the shrub canopy were the result of single-seed caches while the larger caches were placed away from the canopy. Even including the high mortality of single seedlings, the antelope bitterbrush understory in this pine woodland was regenerated by the establishment of both single seedlings and multiple seedlings.

R. L. Sherman and W. W. Chilcote examined bitterbrush seedling clumps in a pine woodland/bitterbrush community in central Oregon and found that the majority of younger bitterbrush clumps were out in the open where litter was sparse. The older clumps were located closer to the canopies and in denser litter.[24] The authors suggested that the reduction in fire frequency at that site has changed the caching behavior of the rodents. The litter that accumulates in the absence of periodic wildfires provides fuel loads sufficient to destroy antelope bitterbrush seedlings. Rodents in such areas selected cache sites more favorable to bitterbrush seedling survival (less litter), and as the intervals between fires became progressively longer, caching was limited to areas farther from individual trees. Areas with more frequent fires and concomitant litter removal had many favorable caching sites. The authors suggested that although individual bitterbrush plants are not fire resistant, bitterbrush is a fire-dependent species because of rodent seed-caching relations.

Earlier in this chapter we discussed the importance of beta-carotene to rodents. Evans et al. were the first to report that rodents cache antelope bitterbrush seeds in the summer and return to graze the cotyledons of the seedlings the following spring when they are rich in carotene. We investigated the influence of rodent predation on antelope bitterbrush seedlings at the Doyle Wildlife Management Area using portable rodent enclosures.[25] Seedlings protected from rodent predation showed 87 percent survivability (survivability = first true leaf stage), whereas seedlings subjected to rodent predation had only a 47 percent chance of surviving. We also studied the effects of different rodent species. Ord's kangaroo rats preyed on 56 percent of the seedlings, followed by Great Basin pocket mice at 25 percent, and deer mice at 21 percent. Kangaroo rats grazing on cotyledons produced an effect similar in appearance to a combine passing through a wheat field. The pocket mice and deer mice took a different approach, combining grazing on the cotyledons with digging for seeds (germinated and ungerminated), thus causing the death of seedlings through soil disturbance. Explosions in meadow mice (*Microtus montanus*) populations have been reported to result in significant antelope bitterbrush seedling predation as well.[26]

Obviously, the interactions between granivores and *Purshia* seeds and their effect on seedling establishment are extremely complex. We have virtually ignored desert bitterbrush and cliffrose in this discussion because little has been reported about the seedbed ecology of these species. We assume that granivore interactions with seeds of desert bitterbrush would be similar to those observed for antelope bitterbrush, although the species composition of the rodent communities would be different. Cliffrose achenes, with their persistent style, may present particular problems to the granivores that collect and cache them.

Chapter Eight

Ruminant Nutrition

The *Purshia* species are very important in the nutrition of mule deer and, to a lesser extent, domestic livestock in the western United States. In order to appreciate the nutritional aspects of bitterbrush, some understanding of the unique nutritional requirements of mule deer, the primary consumers of *Purshia* browse, is necessary.[1]

Analyses of forage samples were among the first scientific tests applied by resource managers attempting to understand the role of forage plants in the ecology of western ranges.[2] Of course, most of these analyses were of grasses rather than shrubs, and the object was to understand the forage requirements of domestic livestock rather than mule deer. By the 1930s, however, ecologists were making serious attempts to understand the nutritional requirements of North American deer.[3]

L. A. Stoddard and J. E. Greaves's analysis of summer ranges in Utah, published in 1942, is one of the classic forage nutrition studies.[4] Although their work was done on a summer range rather than a mule deer winter range, the authors did analyze antelope bitterbrush browse and arrived at a crude protein content of 15.4 percent. They concluded that antelope bitterbrush is a highly reliable source of browse.

Protein is often considered the most important dietary nutrient; without it the body cannot maintain itself. Even a slight deficiency adversely affects reproduction, lactation, and growth. Ruminants need protein in order for the rumen microorganisms to digest and metabolize carbohydrates and fats effectively. This nutrient, usually reported as crude protein, is a measure of protein and nonprotein nitrogen multiplied by a correction factor, usually 6.25.

The amount of nitrogen compounds present in trees and shrubs varies with the kind of tissue, the age or stage of development, and the season. In general, shrubs contain higher percentages of crude protein than do grasses and forbs during fall and winter, but lesser amounts during spring and summer. The leaves of shrubs contain higher percentages of crude protein than the stems, and the tips of the stems contain higher protein levels than the thicker middle and butt sections.[5]

Most of the plant material eaten by deer and other ruminants, however, consists of some form of carbohydrates, which provide most of the energy and furnish needed bulk in the diet. Complex carbohydrates such as starch and cellulose are the most important polysaccharides in trees and shrubs. Cellulose is the chief constituent of the cell walls of shrubs and forms the plant's support structures. Woody plants transform sugar to starch for storage in the summer and fall, and change starch back to sugar for energy in winter. For ruminants, the form of the sugar present probably makes little difference in nutritional value per se, but it may affect preference, and thus nutritional intake. For many years the Weende system of proximate analysis was used to separate carbohydrates into digestible and nondigestible portions. More modern analysis systems compare the solubility of cell contents and cell wall constituents in neutral and acid detergents. This appears to relate more closely to actual animal performance. Acid detergent fiber (ADF) and acid detergent lignin (ADL) and cellulose are often used as indicators of the carbohydrate value of feeds for ruminants.

Fats, lipids, and related substances are frequently reported in the results of proximate analyses of feed stocks as "crude fat" or "ether extract." The term *ether extract* comes from the analytical procedure by which the fats are dissolved or extracted from the sample using ether as a solvent. The true lipids are simple lipids, true fats and oils, compound lipids, and derived lipids such as saturated and unsaturated fatty acids. Most of these are digestible to some degree by ruminants. Some other ether-soluble compounds, such as terpenes and resins, are not digestible and may even be harmful to rumen function.[6] Fats are a highly important food reserve in animals because they contain almost twice as much energy per unit of weight as carbohydrates or protein.

While the fat content may be as high as 70 percent in dry fruits and seeds of shrubs, lipids rarely make up more than 5 percent of the stem and leaf constituents. Fats are synthesized in the rumen from carbohydrates and proteins, so ruminants do not require fats in their diet, but wintering animals prefer and do well on diets consisting of shrubs such as winterfat (*Krascheninnikovia lanata*). The foliage of Rocky Mountain juniper (*Juniperus scopulorum*) tests high for crude fat, but essential oils that interfere with rumen function make up a significant portion of the ether extract.

Almost all of the mineral content of plants can be recovered in the ash left after ignition at 1,240°F. The ash content of forages is important because it provides an indication of digestibility. The two most important minerals normally reported in routine feed analysis are calcium and phosphorus. Other minerals such as sodium, potassium, chlorine, magnesium, iron, sulfur, iodine, manganese, copper, cobalt, and zinc are necessary for many animal body processes but are usually supplied in adequate amounts in common shrub diets. Some of these minerals function as constituents or activators of enzymes and are needed in very small (trace) amounts.

Ruminants must have access to adequate calcium. Calcium and phosphorus compounds form 90 percent of the mineral matter in the skeleton of most ruminants and about 75 percent of that in the entire body. On the western ranges in the United States, calcium supplies are usually ample in shrubs, and in some cases are high enough to adversely affect phosphorus metabolism.

Phosphorus, vital in many body processes, is an essential part of the skeleton, intracellular fluids, and compounds such as nucleoproteins and phospholipids. It is necessary for transfer of energy through the action of adenosine triphosphate (ATP). A deficiency of phosphorus or a high calcium-to-phosphorus ratio may cause retarded growth, weak young, decreased lactation, failure to conceive, and many other abnormalities. A desirable calcium-to-phosphorous ratio is somewhere between 1:2 and 2:1. Phosphorus is deficient in many shrub species on ranges during the winter months. Shrub species that maintain adequate phosphorus levels during the dormant season should be encouraged on winter ranges through management techniques.

Vitamins are organic compounds essential for normal functioning of the body that are usually required in trace amounts. Their chemical role is largely catalytic. They usually form part of or act in conjunction with enzymes. The oil-soluble vitamins (A, D, and E) are a necessary part of the diet of ruminants. The water-soluble vitamins (C, B complex, K, etc.) can be synthesized from other food constituents by rumen bacteria.

Vitamin A is the most important vitamin to range ruminants because it is the most likely to be deficient. Only minute amounts are required, but animals can store it only for short periods. Beta-carotene is the major source of vitamin A in shrubs. Approximately 50 percent of the beta-carotene in forage is converted to vitamin A by the animal body. Shrubs are usually high in carotene during early growth, especially the developing leaves, but levels decline rapidly after maturation and dormant leaves are often deficient. Because evergreen shrubs tend to hold carotene levels higher during the winter, they are a valuable source of vitamin A on winter ranges.

Energy content is an important measure of the nutritive value of feeds because it provides a common basis for expressing nutritive value. Energy values are used extensively in determining the feeding value of shrubs on ranges. With the possible exception of protein or phosphorus deficiencies, the most common problem with mule deer ranges, especially in winter, is lack of available energy. Gross energy levels obtained through analysis can be misleading if they reflect high essential oil contents. These nondigestible oils give off considerable heat upon combustion, indicating high energy values that are not, however, usable by animals.

Most of the ruminant's nutrient intake goes toward maintaining its general metabolism. Browse species are generally good sources of energy, but some are decidedly lower in energy-furnishing constituents than others. Deciduous shrubs are generally lower in energy value than plants that retain leaves during most of the winter. That is why winter leaf retention is such an important feature in certain ecotypes of antelope bitterbrush.

Although proximate analyses of shrubs indicate their probable nutritive value, only feeding trials can provide a definitive analysis of digestibility. Digestion trials have shown, for instance, that mature fall and winter shrub material is less readily digested than tender spring growth. The classic digestion trials involve feeding known amounts of browse or other forage to animals caged or fitted with a collection apparatus. Both the feed and the feces are then chemically analyzed to determine digestible protein and total digestible nutrients. This method is time-consuming and expensive, and collecting sufficient data for a statistically valid test is very difficult. The use of the in vitro laboratory technique has greatly simplified the determination of digestible dry matter. This technique involves the use of an artificial rumen made of tubes and rumen fluid obtained from donor animals. The advantage of this procedure is that large numbers of samples can be digested simultaneously.

Much of what is known about the nutrition of mule deer comes from the pioneering research of Arthur D. Smith, a longtime professor at Utah State University who worked in cooperation with the Utah Fish and Game Commission. Smith's many students have gone on to other universities and agencies and have had a profound influence on wild ruminant nutrition studies.

One of the first papers Smith published on mule deer nutrition concerned the digestibility of sagebrush.[7] Based on field observations, he reported that mule deer commonly consumed big sagebrush herbage. On some winter ranges they consumed large amounts, especially when more desirable species such as antelope bitterbrush were not available. When he maintained two mule deer on a diet restricted to big sagebrush (*Artemisia tridentata*) for two months, the animals did not maintain their body weight. Smith also found significant differences among animals with regard to their preference for sagebrush browse.

Smith next compared the digestibility of curlleaf mountain mahogany (*Cercocarpus ledifolius*), antelope bitterbrush, and Utah juniper (*Juniperus osteosperma*) using captive mule deer.[8] He used alfalfa (*Medicago sativa*) hay as the standard forage because, as a common feed for domestic livestock, a great deal was known about its digestibility. Alfalfa hay was included as well because it was the first choice for artificial winter feeding of mule deer. Of the three native plants, curlleaf mountain mahogany was the preferred, most nutritious, and most digestible forage. The captive deer did not prefer Utah juniper foliage. Antelope bitterbrush provided the least digestible nutrients, but the accession used had virtually no persistent leaves.

Smith also addressed the problem of why wild mule deer fed alfalfa hay during the winter were not surviving. He fed his captive mule deer screened alfalfa hay that consisted largely of leaves. They readily ate this ration and performed well on the diet. The problem with artificially fed wild mule deer, Smith believed, was that they would not consume the stems, which generally constitute the bulk of alfalfa hay. The deer would ignore the stems and search through baled alfalfa for the leaves, which were not present in sufficient quantity to provide adequate forage. At the time, many wildlife managers thought there was something nutritionally wrong with alfalfa for deer. In a sense there was, but not in terms of a toxic material.

Before Smith's work, Everett R. Doman and D. I. Rasmussen of the Utah Department of Fish and Game and the Utah Cooperative Wildlife Research Unit had published on the inherent problems of artificially feeding mule deer and elk in 1944. The Utah winter feeding program had started in 1931 after reports of elk damage to private property. Experience with supplemental feeding of mule deer indicated that at best the programs were only partially successful. The authors considered fine-stemmed alfalfa hay to be the best supplemental forage.

In another publication Smith reported on digestion trials and chemical analysis of mountain mahogany (*Cercocarpus montanus*), cliffrose, chokecherry (*Prunus virginiana* var. *melanocarpa*), and Gamble oak (*Quercus gambelii*), which furnished fair to poor browse during the winter (Table 8.1).[9] Smith's analyses were the first total energy determinations for the browse of the shrubs.

Arthur Smith's pioneering research was not confined to winter ranges; he also conducted preference trials with captive mule deer on summer forage.[10] Until late in May, mule deer consumed a greater amount of herbaceous vegetation than browse. Thereafter, browse became more important in the diet, continuing to increase until the end of the summer, when herbaceous species formed less than 10 percent of the daily consumption and browse plants made up 87 percent. The average consumption of air-dry material was 3.49 pounds per 100 pounds of mule deer weight. Antelope bitterbrush was relatively low on the consumption list, especially during early periods in the feeding trial, which generally agreed with field observations of mule deer diets.[11]

At about the same time that Arthur Smith was working with captive mule deer in Utah, Herbert Hagen was studying deer forage plants in the Sierra Nevada and Harold Bissell and his associates in the California Department of Fish and Game were conducting digestibility trials in California.[12] Bissell et al. used mule deer for digestion trials with antelope bitterbrush and sagebrush, and black-tailed deer (*Odocoileus hemionus* subsp. *columbianus*) in their other forage studies. The mule deer were trapped on antelope bitterbrush ranges in Lassen County, California, during the winter of 1953–1954. Their daily intake of dry bitterbrush was 2.2 pounds per 100 pounds of body weight. Intake was the same in trials conducted in December and in the spring.

TABLE 8.1.
Digestible Nutrients (%) of Browse Plants Compared with Nutrients for Common Livestock Feeds

Species	Protein	Total Digestible Nutrients
Sagebrush	7.3	78.1
Common millet hay	8.2	68.9
Curlleaf mahogany	6.0	65.5
Timothy hay	6.3	56.6
Utah juniper	1.0	63.5
Milo stover	1.2	53.6
Mountain mahogany	3.5	49.6
Field pea straw	3.5	57.4
Cliffrose	3.3	31.2
Sudangrass straw	3.6	49.3
Antelope bitterbrush	2.7	29.6
Bunchgrass hay	2.9	53.1
Chokecherry	0.6	27.3

Adapted from Arthur D. Smith, "Nutritive Value of Some Browse Plants in Winter," *J. Range Manage.* 10 (1957): 162–164.

Arthur Smith, in another study, calculated an average daily consumption per mule deer (not per 100 pounds) of 2.6 pounds.[13] He fed antelope bitterbrush for 12 days to two mule deer and found that the average consumption of that shrub was about 1 pound per day. He obtained similar results with cliffrose. In the same study Smith discovered that mule deer would eat the browse of some juniper and sagebrush plants but rejected the browse of similar-appearing plants of the same species.

Bissell et al. tried feeding a local sagebrush to the same deer, but their intake was too low to maintain body weight. At the time the experiment was conducted, the subspecies of big sagebrush were not generally recognized, and it is not clear what type of big sagebrush was used. The authors calculated the energy requirements for deer from these experiments (it is not clear if the figures are for mule deer or black-tailed deer) and found that a deer consumed about 4,500 calories per 100 pounds of body weight per day. The maintenance requirement for a resting, fasting deer was 1,140 calories per day. Bissell et al. considered all the browse species they tested to contain potentially adequate sources of digestible energy; however, they concluded that antelope bitterbrush was the only plant studied that could be eaten as a sole item in the diet for a prolonged period. The total digestible nutrient level they reported for antelope bitterbrush was 54.8 percent, much higher than that previously reported by Arthur Smith.

R. W. Lassen et al. obtained similar results on the nutritional importance of antelope bitterbrush in their studies of the Doyle, California, mule deer herd.[14] They analyzed the stomach contents from 213 mule deer (including a remarkable 186 road kills; at that time sparsely traveled U.S. 395 by Doyle must have looked like a bumper-car track in an amusement park). The herd's principal browse was big sagebrush; however, many of the mule deer examined were in poor condition and evidently had been starving when killed. In January 1950 the mule deer appeared to be in fair condition. By April 1950 some 30 percent of the deer examined exhibited signs of advanced malnutrition. The herbaceous species most frequently found in mule deer stomachs collected during the *winter* was alfalfa; the second most frequent was cheatgrass (*Bromus tectorum*). Grass constituted 56 percent of the herd's diet from November through March. Once the antelope bitterbrush began its annual growth in April, however, grasses declined to 5 percent of the diet. During the great population crash of the Lassen herd in the early 1950s the herd was dependent on herbaceous species for a large part of its diet. At the same time, mule deer habitat restoration projects concentrated on establishing antelope bitterbrush and excluded herbaceous species from the restoration planting mixes (see Chapter 6). It is very important to note that big sagebrush constituted the major proportion of browse in the mule deer stomachs Lassen examined (Fig. 8.1), although the authors did not specify the species or subspecies of *Artemisia* found. If some forms of the big sagebrush that occurred on the winter ranges were more preferred and more digestible than others, those forms were disproportionately browsed by mule deer, and individual plants may have been as severely grazed as antelope bitterbrush. The significant question then becomes, was the overutilization of sufficient intensity and duration to limit the frequency of preferred forms of big sagebrush in the winter range plant communities?

In a paper published in 1955, Bissell and Strong reported on variations in crude protein in the browse diet of California deer.[15] This is a remarkable publication because of the enormous amount of tabular data it presents, all assembled by H. R. Leach. One of these tables was based on analyses of stomach contents of animals collected from the Lassen-Washoe interstate mule deer herd from September through May (Table 8.2). The plant species composition recorded from the stomach contents reflects a considerable altitudinal variation on the eastern slope of the Sierra Nevada. Squaw carpet (*Ceanothus prostratus*) and snowbush (*Symphoricarpos longiflorus*) are species of higher elevations

Fig. 8.1. Mixed community of mountain big sagebrush and antelope bitterbrush growing at Granite Peak in western Nevada. Many resource managers underestimate the importance of big sagebrush in the midwinter diet of mule deer.

TABLE 8.2.

Stomach Content Analysis (%) for the Lassen-Washoe Counties Mule Deer Herd

Species[1]	Sept.	Oct.	Nov.	Dec.	Jan.	Feb.	Mar.	Apr.	May	Mean
Western juniper	0	0	T	1.2	T	0.8	0.5	0.3	0	0.3
California black oak[2]	6.2	8.0	2.6	T	0.3	0.3	0.3	T	0	2.0
Curlleaf mahogany	10.5	0.9	0.7	5.0	1.6	0.1	T	9.2	23.9	5.8
Chokecherry	3.3	2.2	0	0	0	0	2.3	T	T	0.9
Antelope bitterbrush	60.2	62.3	51.6	20.5	10.0	9.9	0	4.8	14.7	26.0
Squaw carpet	5.8	0.2	0.8	3.9	0	1.1	0.6	2.0	0.3	1.6
Snowbush	T	1.5	3.6	2.6	0.3	0.2	0.2	2.5	0	1.2
Sagebrush	0	3.5	11.8	26.9	64.9	53.4	54.5	34.4	44.4	32.6
Other browse species	6.5	7.8	8.6	3.0	2.1	6.5	2.5	4.4	0	4.6
Total	92.5	86.4	79.7	63.1	79.2	72.3	60.9	57.6	83.3	75.0
Forbs	7.5	12.6	5.9	8.7	0.6	6.8	8.8	8.0	7.5	7.4
Grasses (dry)	T	T	8.1	6.3	16.6	7.7	13.1	7.5	T	6.6
Grasses (green)	T	1.0	6.3	21.9	3.6	13.2	17.2	26.9	9.2	11.0
Total grasses	T	1.0	14.4	28.2	20.2	20.9	30.3	34.4	9.2	17.6

[1] Scientific names: *Juniperus occidentalis, Quercus kelloggii, Cercocarpus ledifolius, Prunus virginiana* var. *demissa, Purshia tridentata, Ceanothus prostratus, Symphoricarpos longiflorus, Artemisia tridentata.*

[2] The article did not indicate whether the California black oak found in the stomach contents during September and October was browse or mast. The black oak acorns disperse during September.

Data collected in the early 1950s by H. R. Leach and reported in H. D. Bissell and H. Strong, "The Crude Protein Variations in the Browse Diet of California Deer," *Calif. Fish and Game* 41 (1955): 145–155.

found in pine woodlands. Chokecherry occurs in riparian environments extending down from the mountain escarpment. Big sagebrush and low sagebrush (*Artemisia arbuscula*) are normally found in openings in the pine woodlands down through the foothills to the upper portions of the basin lake plains, but the subspecies of big sagebrush differ dramatically in morphology, secondary phenolic compound content, and preference exhibited by mule deer along this elevational and environmental gradient. Mule deer are very mobile animals who can move up and down the gradient as they seek forage and browse to fill their nutritional needs, but when deep, crusted snowfalls occur at higher elevations, the available range becomes restricted to the lower elevations.

From September through November, antelope bitterbrush was the dominant browse in the diet of the Lassen-Washoe mule deer herd (see Table 8.2). In December, antelope bitterbrush and sagebrush were of roughly equal importance. From January through May, sagebrush was the dominant browse. Antelope bitterbrush was not important in the spring until May. The antelope bitterbrush and big sagebrush browse that is available from January through March represents a much more restricted population compared with that of September through November. The basin big sagebrush (*Artemisia tridentata* subsp. *tridentata*) that is available during midwinter is generally considered to be the least preferred and digestible subspecies of big sagebrush. Whether or not there are similar significant differences in the nutritional characteristics of the antelope bitterbrush populations from the elevation of the pine woodlands down to the lake plain winter ranges is not known. The relatively high proportion of green grass in the mule deer's diets during December indicates that the data were gathered during a relatively mild winter when there was excellent fall germination of cheatgrass. In Leach's earlier reports on the Lassen herd's diet, he indicated that alfalfa was a significant part of the winter diet; however, alfalfa is not mentioned in this report. There is an inherent bias in using mule deer stomach contents to determine diet composition because the coarser portions of browse species accumulate in the stomach over a period of days while fragments of more succulent portions of herbaceous species pass rapidly through.

Earlier field studies of mule deer's preference for individual forage species involved observing the time an individual deer spent feeding on a specific plant. In a very innovative study, Arthur Smith and Richard Hubbard combined observations of minutes of browsing with determinations of actual browse removal.[16] They found no relationship between the time spent feeding and the amount of browse consumed, and suggested that the latter was the better parameter for determining mule deer diets. Obtaining figures for amount of browse consumed required first measuring and tagging a large number of twigs and then returning to measure again after the plant had been browsed. To get the actual amount or weight of browsed material, unbrowsed twigs had to be measured, cut, dried, and weighed and a relationship between length and weight developed. This allowed the development of statistical relationships between the diameter of the animal-clipped stem and the weight of the material removed. Nothing associated with determining the diet and nutrition of wild animals is ever easy!

Bruce Welch of the USDA, Forest Service, Shrub Science Laboratory at Provo, Utah, has been a major modern contributor to our knowledge of mule deer nutrition.[17] The following discussion of ruminant physiology is largely abstracted from Welch's publications.

The first stage of digestion is ingestion. In the mouth, food is broken up by mastication and mixed with saliva, which acts as a food softener and a lubricant. The extent to which food is masticated in the mouth varies enormously from species to species. Ruminants thoroughly grind grass by regurgitating and remasticating it. Cattle are poor at grinding up small seeds, whereas mule deer are quite good at it. From the mouth, food moves down the esophagus into the stomach. As ruminants, mule deer have compound stomachs and are dependent on microbial fermentation for digestion.

The ruminant stomach is divided into four compartments: rumen, reticulum, omasum, and abomasum. The rumen is the first and the largest of the compartments. Hastily eaten food is stored in the rumen under warm, moist, slightly acid conditions that are ideal for fermentation. This food is later regurgitated, thoroughly chewed, and swallowed back into the rumen for additional microbial digestion. The macerated-digested food particles, along with the bodies of millions of microorganisms, are forced into the reticulum and pass on to the omasum, where large quantities of water are absorbed, concentrating the macerated-digested food and microbial mass. From the omasum the food and microbial mass are forced by peristaltic action into the abomasum, where true digestion begins.

During the fermentation process volatile fatty acids are formed. These are absorbed directly from the rumen into the bloodstream and constitute the animal's major energy source. During microbial fermentation the 10 essential amino acids needed by the animals are synthesized by the rumen microorganisms from plant protein, urea, and inorganic nitrogen.

To quantify the nutrient requirements of an animal you need standards for comparison. The standard nutritional needs of domestic livestock have been determined through experimentation and are available in published reference guides.[18] Such standards are not available for mule deer, but the requirements for domestic sheep are considered to be similar.[19]

The energy needs of animals are expressed in several forms, including total digestible nutrients, digestible energy, and metabolizable energy. Obviously, it is necessary to understand these terms and their abbreviations in order to interpret discussions of the subject. Total digestible nutrients, or TDN, is the sum of all the digestible organic compounds (proteins, sugar, cellulose, etc.), with the digestible crude fat component being multiplied by the heat factor 2.25.[20] The TDN requirement of an animal is expressed as kilograms per animal per day or as a percentage of the diet. Digestible energy (DE) is calculated by subtracting the gross calories in the feed from the calories in the feces. The DE requirement of an animal is expressed as megacalories (C) per animal per day or as megacalories per kilogram of dry matter.

The energy needs of mule deer vary according to the animal's weight and activity. Larger animals require more TDN per day for a given activity than do smaller animals. A lactating female requires more TDN per day than a nonlactating female of similar weight. On a constant weight basis, lactation is followed in descending order of energy required by fattening, growth, gestation, and maintenance.

In contrast to the generally held opinion that antelope bitterbrush is an excellent or even essential winter forage for mule deer, Bruce Welch suggested that it is more important in late fall. For evidence that deer do not prefer antelope bitterbrush as a midwinter browse he cited Leach's work in northeastern California, Paul Tueller's in Nevada, and the unpublished work conducted by D. E. Medin in Colorado.[21] Welch and Andrus reported heavy winter use of big sagebrush and wild rose (*Rosa* spp.) and limited use of antelope bitterbrush.[22] Leach considered the switch from antelope bitterbrush to sagebrush to be independent of the severity of the winter. We have discussed Leach's study with field biologists who were working with the Lassen-Washoe mule deer herd at the time and have something to add to that statement. There were severe winter deaths in these mule deer herds for several winters during the late 1940s and 1950s, and the stomachs of winter killed deer often contained nothing but big sagebrush herbage. Based on this knowledge, generations of mule deer managers have accepted the notion that the animals eat big sagebrush when they are starving and then die. At the time Leach conducted his study, however, most of the antelope bitterbrush was severely overutilized and simply was not available to the deer, which had no choice but to switch to sagebrush. Tueller likewise reported that the switch from antelope bitterbrush to sagebrush on Nevada ranges was usually attributed to overutilization of the antelope bitterbrush.

Leach, who is now retired, remains convinced that mule deer use antelope bitterbrush in the fall and again in the spring when the new leaves appear, but during midwinter voluntarily switch to sagebrush. He bases his conclusions on analyses of more than 3,000 deer stomachs during his career. When he originally developed this opinion, he was sharply criticized by W. P. Dassman and August Hormay, both strong proponents of the concept of *key species* in deer management.[23] For the mule deer herds of northeastern California, they believed the key species to be antelope bitterbrush. Leach describes this initial criticism as being on a professional level only, but when he reported that in mild winters a significant portion of the mule deer's diet was foliage of the alien cheatgrass, it was more than his fellow professionals could stomach.

Galen Burrell, who studied the Entiat mule deer herd in eastern Washington, reported the all too familiar story of reduction in antelope bitterbrush habitat due to wildfires, decadence, and lack of regeneration.[24] Burrell's study site was an antelope bitterbrush/bluebunch wheatgrass area adjacent to ponderosa pine/antelope bitterbrush woodlands. He based his conclusions regarding the winter diet of the mule deer on microscopic analysis of fecal samples.[25] The bulk (87–93 percent, depending on site and year) of the winter diet was made up of antelope bitterbrush, buckwheat (*Eriogonum* spp.), arrowleaf balsamroot (*Balsamorhiza sagittata*), and lupine (*Lupinus* spp.). Burrell also compared the composition of the diet in three areas of critical winter habitat where the density of antelope bitterbrush varied. On one site the antelope bitterbrush had been killed by wildfires. On a second site the stand had been partially burned, and the third site was unburned. The first year of the study, antelope bitterbrush (when available) constituted 70 percent of the deer's diet until March, when it dropped to 17 percent. The second year, antelope bitterbrush constituted 86 percent of the diet through the winter and then dropped to 69 percent in March. In the burned sites the antelope bitterbrush was essentially exhausted by December. Burrell used a selection index method to determine which of the available forage was preferred and found that antelope bitterbrush leaders were highly preferred throughout the winter. Burrell cited Richens's studies in northeastern Utah and Wilkens's in the Bridger Mountains of Montana to support his conclusion that the general decline in use of antelope bitterbrush over the winter is caused by a decline in the availability of leaders for browsing.[26]

Two points stand out in Burrell's study. First, it is obvious that mule deer's winter diets do not consist of a *single* species. Burrell identified 27 plant species in the fecal samples. As the most preferred species declined, the animals compensated by selecting other plants. The second striking point is the lack of sagebrush in the winter diet. It is difficult to visualize an antelope bitterbrush community at the edge of a ponderosa pine woodland without some big sagebrush. Was it not present, or did the deer simply not eat it?

R. H. Hansen and B. L. Dearden used fecal analysis to determine the midwinter diet of mule deer in the Piceance Basin of western Colorado.[27] Mule deer winter kills of considerable magnitude had occurred in the basin for several years, and the samples were collected at the most critical time. The deer were assumed to be under extreme hardship because of snow cover and cold temperatures. Pinyon (*Pinus edulis*) and Utah juniper composed 83 percent of the total foods eaten between December and March; big sagebrush, antelope bitterbrush, and Utah serviceberry (*Amelanchier utahensis*) contributed 13 percent. An additional 10 species of plants were eaten in small quantities. The authors suggested that the unusually high dependence on pinyon and juniper might reflect the especially harsh winter conditions and the browse available on the sites.

It should be obvious that many wildlife biologists grossly underestimate the importance of sagebrush in the midwinter diet of mule deer. We agree with Brian Welch that antelope bitterbrush is an excellent browse for mule deer during the fall. We also agree with Welch's convincing evidence of the shortcomings of antelope bitterbrush as a sole source of nutrition during the winter. He reported that the total digestible nutrient requirement for winter mule deer is close to the requirement for winter maintenance of domestic sheep, which is set by the National Academy of Sciences at 55 percent. Numerous trials of antelope bitterbrush fed to mule deer have given results ranging from 39.7 to 54.8 percent TDN, with an average of about 47.7 percent.[28] Welch made the point that the total digestible nutrients in curlleaf mahogany (64.8 percent), big sagebrush (63.4 percent), and even species of juniper (48.4 percent) are higher than the mean for antelope bitterbrush. Generally, the herbage of evergreen woody plants like mahogany, sagebrush, and juniper is more digestible than the bare twigs of deciduous plants such as antelope bitterbrush. Even dormant grasses tend to be higher in TDN than deciduous shrubs.[29]

Welch made a similar comparison for crude protein. The National Academy of Sciences' protein requirement for wintering sheep is 8.9 percent, which is presumably similar to the requirement for mule deer. Antelope bitterbrush has a reported crude protein content of 7.9 percent, desert bitterbrush 8.0 percent, and cliffrose 8.4 percent. Most of the evergreen shrubs such as big sagebrush contain in excess of 10 percent crude protein.[30]

O. Eugene Hickman conducted detailed studies of the nutritive content of range forage species growing near Silver Lake, Oregon, the home range of a famous mule deer herd.[31] His study is important because he followed the nutritive trends in forbs, grasses, and shrubs, including antelope bitterbrush, throughout a growing season. Using a much lower crude protein content (about 6.5 percent) as the minimum for mule deer than that used by Welch, Hickman found the crude protein level of antelope bitterbrush to be adequate during all months of the year.

A more recent study by Carl Wambolt et al. provides a very interesting insight into antelope bitterbrush as winter forage for mule deer.[32] The authors found the crude protein level of antelope bitterbrush leaders to vary from 6.5 to 7.2 percent. The presence of leaves greatly increased crude protein levels. There were so few leaves present by February, however, that they increased the crude protein level by only 0.3 percent. February crude protein averaged 6.8 percent for total available browse, which is below the estimated requirement for wintering mule deer. An interesting aspect of this study is the authors' finding of significant differences in leaf and total crude protein among sites and among sampling dates. They designed the study to include antelope bitterbrush from a variety of environmental conditions in a relatively limited geographical area. The site differences in crude protein suggest that the characteristic is at least partially heritable. With an obligate outcrossing species such as antelope bitterbrush, the variation among populations is to be expected.

Bruce Welch measured crude protein content and retention of winter leaves in *Purshia* and *Fallugia* grown in a common garden in Utah (Table 8.3).[33] The data obtained in that study proved a strong correlation between winter leaf retention and crude protein content. During the 1970s J. M. Alderfer noticed that winter leaf retention varied among antelope bitterbrush populations in Oregon.[34] It appears that winter leaf retention is a cliffrose and desert bitterbrush characteristic that may be segregating in some antelope bitterbrush populations (i.e., Lassen, California).

Mule deer's preference for a particular species or ecotype of browse species has been well documented. During the severe winter kills in the Devil's Garden herd in the early 1950s, a famous photograph was taken of a doe standing on her hind legs to browse on western juniper foliage on a well-high-lined tree. In the background are unbrowsed trees with branches reaching the ground. The crude protein of the two ecotypes of western juniper may be the same in chemical analyses, but obviously the deer can distinguish them. And if a starving animal will not browse on the foliage of one type, it is effectively removed from the food supply. The preference mule deer exhibit for a given browse is governed both by the animals' previous experiences with food sources and by inherent differences in the animals themselves. For example, cliffrose planted at Granite Peak, Nevada, far north of the natural range of the species, where native antelope bitterbrush is heavily utilized by mule deer, initially went uneaten. Antelope bitterbrush plants transplanted at the same time were continually hedged, and many were killed by excessive browsing. The cliffrose plants were not browsed for a decade. One winter, after a decade of avoidance, the mule deer started browsing the cliffrose plants. We have noted similar occurrences with fourwing saltbush (*Atriplex canescens*) plantings. Populations of this dioecious shrub often segregate for herbivore preference; some plants will be browsed to the ground while others remain uneaten. This was apparent in a planting at Juniper Hill in western Lassen County, in which some shrub seedlings were killed by browsing and others were not touched until an exceptionally hard winter occurred. Necessity overcame lack of preference then, and the previously unbrowsed plants were hedged back to coarse branches.

TABLE 8.3.
Winter Crude Protein and Leafiness (%) of Rose Family Shrubs Grown in a Utah Common Garden

Accession	Origin (County)	Crude Protein	Winter Leaves
Apache plume	Sevier, UT	4.8a[1]	—
Antelope bitterbrush	Moffat, CO	5.9a	7.4ab
	Juab, UT	6.6b	13.1bc
	Carbon, UT	6.8b	9.0a-c
	Ada, ID	6.9b	5.9a
	Lassen, CA	7.9c	15.1c
Desert bitterbrush	Washington, UT	8.6cd	49.5e
	Mono, CA	9.3d	50.5e
Cliffrose	Utah, UT	8.8cd	47.5e

[1]Means within columns followed by the same letter are not significantly different at the 0.05 level of probability.

Adapted from Bruce L. Welch, Stephen B. Monsen, and Nancy L. Shaw, "Nutritive Value of Antelope and Desert Bitterbrush," in *Proceedings of the Symposium on Research and Management of Bitterbrush and Cliffrose in Western North America,* ed. A. R. Tiedemann and K. L. Johnson, 173–175 (Gen. Tech. Rep. 152, USDA, Forest Serv., Ogden, Utah, 1983).

Preference is also a function of the intensity of competition for the browse resource and the abundance and nature of alternative forage sources. We noted a subtle example of preference interactions in a reciprocal garden experiment we conducted with big sagebrush plants.[35] We had five gardens, all containing reciprocal plantings of the local collections. Mountain big sagebrush (*Artemisia tridentata* subsp. *vaseyana*) was growing naturally at two of the sites, and basin big sagebrush grew at the other three. Mule deer were abundant during the winter at the two mountain big sagebrush sites. In replicated gardens, the seedlings of the basin big sagebrush plants were never browsed. The mountain big sagebrush plants were not browsed at the gardens where they were native, but they were browsed at the other mountain big sagebrush garden. For example, seedlings native to Granite Peak were browsed at the Churchill Canyon number 5 garden, but not at their native site, and vice versa. It was not a case of a single plant being accidentally grazed; the mule deer searched through the gardens and selectively ate all of the mountain big sagebrush plants to the ground.

When the Devil's Garden herds were crashing in northeastern California and there was no browse available on the overutilized, decadent stands of antelope bitterbrush, unutilized stands of antelope bitterbrush were growing just to the southwest on the flanks of the Medicine Lake Highlands. Unsuccessful attempts were made to lure deer to these stands by distributing salt blocks.[36] We have not looked at all of the antelope bitterbrush stands on the northern flank of the Medicine Lake Highlands, but plants sampled from the roadside leading down to Lava Beds National Monument from Medicine Lake are strikingly different from the Devil's Garden plants. The leaves of the Medicine Lake material are vividly green, not pubescent gray-green, and are viscid to the touch. It seems likely that the ungrazed stands existed because mule deer did not prefer their browse.

Welch suggested that some plants eaten by mule deer in winter are deficient in phosphorus.[37] The National Academy of Sciences' standard for domestic sheep is 0.24 percent, and the content of antelope and desert bitterbrush is 0.13 and 0.10 percent, respectively. Even the leaves of evergreen species such as big sagebrush and juniper are below the phosphorus requirement level during the winter. Furthermore, Hickman found that the phosphorus and calcium content of antelope bitterbrush browse varied from one plant community to the next.[38] Plants that grew as understory species in ponderosa pine woodlands had higher phosphorus levels than those in open antelope bitterbrush/Idaho fescue communities. Dealy reported similar results for ash content of antelope bitterbrush plants growing in natural and thinned ponderosa pine stands.[39] He observed that mule deer favored the browse in the openings even though there was no difference in crude protein content compared with plants in the natural woodland.

Welch considered all browse species to furnish enough carotene to meet the winter vitamin A requirements for sheep and mule deer. Range animals that depend on dormant grasses over the winter, however, can easily develop a vitamin A deficiency.[40]

"Foods of the Rocky Mountain Mule Deer," by Roland C. Kufeld et al., lists 52 published reports of mule deer foraging on antelope bitterbrush.[41] But mule deer are not the only big game animals that eat *Purshia* browse. Studies in Jackson Hole, Wyoming, have shown that the browse is also used by elk (*Cervus canadensis*), moose (*Alces*), and bighorn sheep (*Ovis canadensis*).[42] To this list can be added the Sierra Nevada bighorn sheep (*Ovis canadensis* subsp. *californiana*) and the pronghorn (*Antilocapra americana*).[43] Another very important herbivore that should never be overlooked in discussions of bitterbrush feeders is the jackrabbit (*Lepus californicus*).[44]

None of these animals relies solely on bitterbrush. Over the course of the twentieth century mule deer have learned to eat alfalfa and have become significant pests in alfalfa fields. According to Dennis Austin and Philip Urness, farmers in southern Utah began reporting alfalfa losses to mule deer in the 1930s.[45] Events in Juniper Hill in western Lassen County illustrate the process. In the 1930s the hill supported a dense stand of antelope bitterbrush and mountain big sagebrush. Western juniper trees that became established in the late nineteenth and early twentieth centuries gradually overtopped the browse species and had outcompeted them by midcentury. The wintering mule deer herd greatly declined at this location, but some of the deer that remained became year-round residents, foraging in the alfalfa fields and using the western juniper woodlands for cover. In the winter, the mule deer supplemented their diet with alfalfa hay from stacks and feed grounds.

Mule deer display considerable variability in their preferences for western juniper trees on Juniper Hill. Some trees have been high-lined to the extent that no foliage is in reach of mule deer, and some have never been browsed. In a juniper utilization experiment we harvested blocks of these trees for fuel wood. The alfalfa field mule deer fed on the limb slash from the preferred trees with a vengeance the first night the trees were cut. They might have been alfalfa field deer, but they had a craving for browse when it suddenly became available. The extensive feeding on fresh juniper foliage produced mass digestive upsets in the resident mule deer, to the point that the woodcutting crew refused to work in the area the next day. The same craving for browse was apparent when we transplanted browse species into cleared areas of woodland.

Welch summarized his review of the nutrition value of *Purshia* by saying that antelope bitterbrush supplies more carotene, crude protein, and phosphorus, but less energy, than dormant grass. Desert bitterbrush and cliffrose supply a little more crude protein and energy than does antelope bitterbrush. Big sagebrush, fourwing saltbush, and curlleaf mahogany supply significantly more crude protein, phosphorus, and energy than does antelope bitterbrush. Welch suggested that the deer's gradual dietary switch from antelope bitterbrush to big sagebrush during the transition from fall to winter could be the result of a need for more nutrients in response to increasing environmental stress. Other authors have reported similar or strongly opposed results and conclusions. We believe the inherent variability among *Purshia* populations, differences in the amount and species composition of forage and browse available, and inherent differences among mule deer herds may all contribute to these variable results. It is safe to conclude, however, that in the fall, mule deer depend on antelope bitterbrush browse if it is available. During midwinter many herds depend on big sagebrush. Antelope bitterbrush is a highly preferred browse for mule deer at all seasons except midwinter and early spring before bud break.

Chapter Nine

Insects and Plant Diseases

Compared with plants, soils, and vertebrate animals, insects do not receive a lot of attention from the average natural resource manager or scientist. Few people doubt the importance of insects in wildland ecosystems, but attempts at in-depth understanding of their roles in communities are frustrated by the insects' exceptionally complex taxonomy and life cycles. Juveniles and adults are often entirely different in both appearance and diet. An immature insect observed damaging a range plant may be impossible to identify because the only description of the species is based on the adult stage. Rearing immature insects to the adult stage so that positive identification can be made is often a difficult or fruitless undertaking. Even trained entomologists tend to specialize in one type of insect to lessen the identification load. Luckily, several noted entomologists have recognized the importance of *Purshia* species in rangeland ecosystems and have compiled lists of insects associated with them, and in some cases have done more detailed studies of specific insects.

Their abundant and fragrant flowers make antelope bitterbrush plants especially attractive to insects.[1] Malcolm Furniss offered an eloquent description of insect–antelope bitterbrush relations:

> Many [insects] are phytophagous in their habit, feeding on foliage, sucking juice, or infesting the bark or roots. Others are probably just perchers or somewhat chance visitors, perhaps attracted by the flowers; for example, parasitic wasps that are kept alive by nectar until various insect hosts become available. Doubtless, many insects spread pollen from flower to flower in their travels, including flies and beetles besides the fabled bees. Others, such as spiders and predacious insects, lurk or hunt for prey amongst the canopy.[2]

Furniss also pointed out that insect sampling tends to be highly biased toward sites where natural resource research projects are located or where destructive outbreaks of a certain insect have occurred. His original study of antelope bitterbrush insects, published in 1972, lists 81 insect species associated with the shrub. The number had grown to more than 100 by 1982 when Furniss addressed a symposium on bitterbrush and cliffrose. Of these, 76 species were phytophagous:

Order	*Family*	*Species*
Coleoptera	3	8
Diptera	1	3
Hemiptera	6	20
Homoptera	7	19
Lepidoptera	10	20
Orthoptera	1	5
Thysanoptera	1	1

For a complete list, see Table 9.1.

Although Furniss found caterpillars (Lepidoptera) and sap-sucking insects (Hemiptera, Homoptera) to constitute the most numerous insect species collected from antelope bitterbrush, this does not preclude the other orders from having equal or greater importance in bitterbrush ecology.

Insect injury has long been suspected to be the leading cause of bitterbrush seed damage. In Chapter 6 we noted that every collection of antelope bitterbrush seeds contains at least some, and sometimes many, shriveled black or spotted seeds. In the 1960s the Forest Service research unit at Boise, Idaho, launched a study of the effects of insects on the flower buds and fruits of antelope bitterbrush.[3] This research group selected 600 flower buds for weekly observation, from seed production through seed dispersal, and reported the following distribution of seed injury:

Stage	*Causal organism*	*Percentage of total*
Normal seeds	—	25.0
Buds and flowers	thrips and unknown	6.5
Fruits	gelechiid caterpillars	5.0
Seed	midge	6.8
	unknown insect	0.7
	Chlorochroa sayi	11.2
	spotted	48.8

Insects damaged 39 of the 600 (6.5 percent) selected buds while they were in the expanding bud, or early flowering, stage.[4] Most of the bud damage was attributed to the narrow-winged thrips, *Frankliniella occidentalis*. The thrips gained entry by chewing through the side of the bud, and once inside fed on the reproductive organs.

A major source of insect damage to antelope bitterbrush seeds in the Idaho studies was a seed midge (according to Furniss it was a species of *Mayetiola*; Basile et al. considered it to be a species of *Phytophaga*; and Ferguson et al. were sure the insect was *Frankliniella occidentalis*).[5] The adult midge, a small,

TABLE 9.1.
Insects Collected from Purshia tridentata

ORDER Family Genus and species	Part of plant affected
COLEOPTERA	
Buprestidae	
Acmaeodera purshiae (Fisher)	Stem
Chrysobothris deleta (LeC.)	Stem
Chrysomelidae	
Altica bimarginata (Say)	—[1]
Cryptocephalus sanguinicollis (Suffrian)	—
Chrysomela lineattopunctata (Forster)	—
Monoxia sp. poss. *consputa* (LeC.)	—
Pachybrachis sp.	—
Curculionidae	
Triglyphulus sp.	—
DIPTERA	
Cecidomyiidae	
Dasineura sp.	—
Leucopis sp.	—
Mayetiola sp.	Fruit
HEMIPTERA	
Coccidae	
Platypedia sp.	Stems
Coreidae	
Harmostes reflexulus (Say)	—
Lygaeidae	
Nysius angustatus (Uhler)	—
Geocoris pallens (Stal)	—
Miridae	
Adelphocoris rapidus (Say)	—
Atractotomus purshiae (Froeschner)	Leaves
Capsus simulan (Stal)	—
Ceratocapsus sp.	—
Deraecoris fulgidus (V.D.)	—
Labops hirtus (Knight)	—
Osallus pilosulus (Uhler)	—

TABLE 9.1.
(continued)

ORDER	
Family Genus and species	Part of plant affected
Pentatomidae	
Apateticus crocatus (Uhler)	—
Brochymena quadripustulata (Fab.)	—
Holcostethus abbreviatus (Uhler)	Leaves
Chlorochroa ligata (Say) *sayi* (Stal)	Leaves, seed
Thanta pallidovvirens (Stal)	Leaves
Zicrona caerulea (Linn.)	—
Scutelleridae	
Eruuygaster sp.	—
Homaemus bijugis (Uhler)	—
Tigidae	
Gargaphia opacula (Uhler)	Leaves
HOMOPTERA	
Aphididae	
Macrosiphum avenae (Fab.) *purshiae* (Palmer)	—
Cercopidae	
Aphrophora permutata (Uhler)	—
Cicadellidae	
Asceratagallia sp.	—
Erthroneura sp.	—
Gyponana sp.	—
Neocoelidia sp.	—
Osornellus borealis (DeLong) and (Mohr)	—
Paraphlepsius sp.	—
Scaphytopius sp.	—
Cicadidae	
Anisococcus quercus (Ehrhorn)	—
Lecanium cerasifex (Fitch)	Stem
Phenacoccus eriogoni (Ferris)	—
Diaspididae	
Lepidosaphes ulmi (Linn.)	Stem

TABLE 9.1.
(*continued*)

ORDER Family Genus and species	Part of plant affected
Psyllidae	
Arytaina pubescens (Crawford)	—
Psykka coryli (Patch)	Fruit
Psykka hirsuta (Tuthill)	—
Psykka media (Tuthill)	Leaves
LEPIDOPTERA	
Crambidae	
Crambus plumbifimbriellus (Dyar)	—
Gelechiidae	
Filatima sperryi (Clark)	Leaves
Filatima sp. (near *abactella*)	Leaves
Gelechia mandella (Busck.)	Leaves
Geometridae	
Amacamptodes clivinara profanta (B.& McD.)	Leaves
Chlorosea sp. (prob. *margaretaria* Sperry)	Leaves
Chlorosea nevadaria (Pack.)	Leaves
Itame colata (Grt.)	Leaves
Marmpoteryx marmorata (Pack.)	Leaves
Semiothera californiaria (Pack.)	—
denticulata sexpuncata (Bates)	Leaves
Lasiocampidae	
Malacosoma californicum (Pack.)	Leaves
Lymantriidae	
Orgyia vetusta gulosa (Boisd.)	Leaves
Psychidae	
Apterona helix (Siebold)	Leaves
Pyralidae	
Ephestiodes erythrella (Ragonot)	Leaves
Myelopsis coniella (Ragonot)	Leaves
Saturniidae	
Hemileuca nuttalli (Strecker)	Leaves

TABLE 9.1.
(*continued*)

ORDER	
Family Genus and species	Part of plant affected
Tortricidae	
Choristoneura rosaceana (Harr.)	—
Sparganothis tunicana (Wism.)	Seedlings
Yponomeutidae	
Trachoma walsinghamiella (Bsk.)	Leaves
ORTHOPTERA	
Acrididae	
Melanoplus sanquinipis sanquinipes (Fab.)	—
Melanoplus bivittatus (Say)	—
Oedaleonotus enigma (Scudder)	—
Schistocerca lineata (Scudder)	—
Spharagemon equale (Say)	—
THYSANOPTERA	
Thripidae	
Frankliniella occidentalis (Pergande)	Buds & flowers
ACARI	
Eriophyidae	
Aceria kraftella (K.)	Leaf galls
Aceria tridentatae (K.)	Twig galls
Phytoseiidae	
Typhlodromus mcgregori (Chant)	—
Tetranychidae	
Eotetranychus perplexus (McG.)	Leaves
Bryobia sp.	Leaves

[1] Part of plant affected by insect not specified.

From M. M. Furniss, "A Preliminary List of Insects and Mites That Infest Some Important Browse Plants of Western Big Game" (Res. Note 155, USDA, Forest Serv., Ogden, Utah, 1972).

Fig. 9.1. Say's stinkbug (*Chlorochroa sayi*) preys on antelope bitterbrush seeds by piercing and sucking plant juice.

slender fly similar in appearance to a mosquito, lays its eggs on antelope bitterbrush flowers between the calyx tube and flower bracts in early May. After hatching, up to a dozen orange larvae (Basile et al. described them as pinkish) mine through the flower bud and feed on the developing seed, causing it to shrivel and blacken. The mature larvae form puparia known as the flax-seed stage because of the pupal case's appearance. Infested fruits are easily recognized by their more narrow aspect and because they persist on the plants after leaf drop in winter.[6] Ferguson et al. suggested that the major source of mortality for the midge larvae was parasitism by the eulophid parasitic wasps *Aprostocetus* sp. and *Tetrastichus* sp. The surviving larvae appeared to overwinter within the puparia and complete their development in the spring to coincide their emergence with the formation of flowers and seeds.

Many of the antelope bitterbrush flowers that produced seed exhibited blackened and shriveled achenes or achenes with necrotic spots, suggesting an insect capable of piercing the seed and sucking plant juice. The prime suspect was Say's stinkbug (*Chlorochroa sayi*), which had often been seen feeding on antelope bitterbrush (Fig. 9.1). The tiny openings in the surrounding tissue left by the bug's pricking action are difficult to detect, however. To be certain that the stinkbugs were responsible for the seed damage, Basile and Ferguson conducted caged trials with nylon net sleeves over flower branches with and without the bugs.[7] The cages with stinkbugs produced 1,742 antelope bitterbrush seeds, of which 84 percent were blackened and shriveled. The caged branches without bugs produced 1,944 seeds, of which 2 percent were blackened. Related species, the green plant bug (*Chlorochroa uhleri* Staal) and the conchuela (*C. ligata* Say), are serious pests of antelope bitterbrush seeds in California.[8] Spotted seeds occurred in the cages without stinkbugs, and the authors considered this evidence that the spots are not caused by stinkbugs.

Furniss sent a query to fish and game departments in the western states and provinces of North America asking which insects were of greatest concern to wildlife managers. The most frequent complaint was of the tent caterpillar (*Malacosoma californicum* Packard) on antelope bitterbrush.[9] Furniss considered the importance of the tent caterpillar to be overrated, however, because tent caterpillar infestations in Idaho were usually restricted to individual plants and did not last for more than a year.[10]

The classic studies of the tent caterpillar and antelope bitterbrush were undertaken by Edwin C. Clark of the Biological Control Laboratory of the University of California at Berkeley,[11] apparently with the encouragement of August Hormay. Clark identified the tent caterpillar that was widespread in antelope bitterbrush stands as *Malacosoma fragile,* the Great Basin tent caterpillar, a species first described by R. H. Stretch in 1881 from material collected near Virginia City, Nevada.[12] F. P. Keen gave the general distribution of this insect as between the Sierra Nevada–Cascade and Rocky Mountains.[13] The adult *Malacosoma* lays her eggs in the fall, and the eggs overwinter in that stage. Egg hatch occurs in the spring, and larval development requires 40–70 days. Pupation, adult flight, and oviposition occur in mid-to-late summer. Russel Mitchell, who conducted detailed studies of the life history of *Malacosoma californicum* in the pine woodlands on the eastern flank of the Cascade Mountains in Oregon, reported that egg hatch occurred in late March at 50 degree-days, just at the beginning of bud burst on the exclusive host plants, antelope bitterbrush and squaw currant (*Ribes cereum*).[14] (He calculated the accumulation of degree-days above a threshold of 5.5°C.) Clark suggested the hatch date was related to the formation of flower buds on antelope bitterbrush and varied with latitude, altitude, and seasonal variations in weather. Mitchell observed five instars (*instar* refers to the period between molts) in larvae growth, with the colonies being dispersed in late May (300 degree-days) when antelope bitterbrush was in full bloom. At this time 40 percent of the population was in the fifth stadium. Pupation began about 1 week later at 400 degree-days. Moth flight peaked in early July at 800 degree-days.

Tent caterpillar eggs are deposited in masses or bands containing 100–140 eggs around the smaller stems of the host plant in July through September, held in place with a dark, frothy, cementlike material termed spumaline. Immediately after hatching, the gray-black larvae cluster on the surface of the egg mass. Shortly thereafter, they move to the tip of the twig, where they begin feeding on the young foliage and start constructing the tent. The tent consists of layers of webbing around the tip of a twig. It is small and relatively inconspicuous for the first two to three weeks of development and may be only 1–2 inches long. The tent is continually enlarged as the larvae develop and attains a length of 5–6 inches at larval maturity. In dense populations, larvae from different egg masses may join and construct tents more than a foot long (Fig. 9.2).[15]

During the course of development, the dominant color of the larvae changes from the black or grayish black of the newly hatched caterpillars to yellow during the second and part of the third instars and then to reddish brown and red

Fig. 9.2. Tent caterpillars (*Malacosoma fragile*) infesting antelope bitterbrush.

during the fourth and fifth instars. The caterpillars are highly gregarious during early development, with individuals of a single brood found concentrated on a single twig. Final-instar larvae usually feed individually or in small groups.

Just prior to pupation, the mature larvae, which are 1½–2 inches long, become completely nongregarious. They wander at random over the host or on the ground and may be found several hundred feet from the nearest host plant. As a result of this wandering, the yellow cocoons of *M. fragile* are widely scattered over the infestation site. Although they do occur in twisted leaves, attached to stems or twigs, or in old tents, pupae are found most frequently under layered stems or debris on the ground surface.[16] Adult flight follows the onset of pupation by about two weeks. The adult males are reddish brown and have a wing expanse of about ¾–1 inch. The females are creamy yellow with a wing expanse of 1–1¼ inches. Mating and egg laying occur shortly after the adults emerge from the cocoons.[17]

The Great Basin tent caterpillars Clark studied fed on a number of woody species found growing in association with antelope bitterbrush, including members of the family Rhamnaceae (*Ceanothus*), Salicaceae (*Populus* and *Salix*), and several other members of the Rosaceae (*Rosa, Prunus*). Clark noted that the wandering larvae were not particular about where they fed, but adults were quite specific regarding the plant species where they oviposited, and antelope bitterbrush was the primary host.

The larvae hatch when the first leaves are about half grown. Feeding continues throughout flowering and fruit maturation and through about three-quarters of the period of twig development. Thus, the feeding period of the insect encompasses the period of maximum physiological activity of the antelope bitterbrush plants. At high insect densities, the young caterpillars first destroy the leaves on a portion of the plant and later spread and defoliate the entire plant. Heavy feeding prevents twig growth and may prevent twig elongation. When the caterpillars are finished with it, the plant may be devoid of photosynthetic tissue.[18]

Plants that are completely defoliated are capable of producing new leaves in the same season. Clark reported that in 1953, foliage was completely replaced in September on plants that had been bare in August, and some of the leaves were nearly full size. Twig growth started again as well, but reached only ½–1 inch in length versus more than 5 inches for plants that were never infested. Mitchell was certain that infestations of western tent caterpillars on antelope bitterbrush in Oregon had no long-term effect on the plants, even when they were defoliated in two consecutive seasons.[19] This appears to be a gross oversimplification. Remember that antelope bitterbrush flowers on second-year wood. Reduction or elimination of twig growth must reduce flowering and potential seed production the next season or, in the case of consecutive years of defoliation, for several seasons.

The presence of the tents may deter browsers. We have not observed browsing on infected plants while the caterpillars were active. This, of course, automatically increases browsing pressure on uninfested plants in the same stand. Clark reported Hormay's belief that the leaves regrown after defoliation by Great Basin tent caterpillars are especially succulent in the fall and are subject to severe browsing by mule deer and livestock.[20]

With the benefit of several decades of careful observation, Bill Phillips has suggested that the influence of Great Basin tent caterpillar infestations on antelope bitterbrush may depend on the location and characteristics of the particular bitterbrush community.[21] At higher elevations in pine woodlands, infestations probably have a minimal influence on the sustainability of the stands. At low elevations, where antelope bitterbrush occupies critical mule deer winter range in the sagebrush zone, the influence of the infestation may be catastrophic. Phillips referred to such winter ranges as the "twilight zone" for antelope bitterbrush. The plants are often senescent and overutilized, reproduction is lacking, and recurrent drought is stressing the plants. In such situations, a tent caterpillar infestation may well be the last straw.

Tent caterpillar outbreaks are cyclical. Mitchell suggested that infestations commonly run for four years in central Oregon.[22] Clark used Forest Service records to trace a large infestation that occurred in 1943–1944 in antelope bitterbrush stands in the Devil's Garden area of Modoc County, California, that were considered essential winter range for the Devil's Garden mule deer herd. The herd was probably near its peak in numbers when the infestation occurred, and the antelope bitterbrush community was already heavily overutilized. At its maximum in 1944 the tent caterpillar infestation covered about 70,000 acres. On about one-half of this area the antelope bitterbrush was completely defoliated, and according to Forest Service records most of those plants died.[23]

Tent caterpillar populations usually crash after three or four years of spectacular increases because of natural virus epizootic outbreaks. The populations Mitchell studied in Oregon crashed between dispersal and pupation. Clark and Clarence Thompson suggested that the crashes are due to increases in a polyhedrosid virus.[24] They actually tested artificial inoculation of this virus in Great Basin tent caterpillars in antelope bitterbrush stands near Truckee, California. Some of the tests were quite promising, but not all the caterpillar populations were totally controlled—perhaps, the authors suggested, because of natural variability in susceptibility.[25]

The webs, dry leaves, and waste associated with Great Basin tent caterpillars increase both the chances of ignition and the rate of spread of wildfires. In fact, this is probably one of the major dangers of infestations. Combine dry tent caterpillar waste with a cheatgrass understory on a hot day in August, and the stage is set for a very serious wildfire.

In his study of antelope bitterbrush insects in Idaho, Furniss discovered a different type of caterpillar damaging seed production.[26] Two species of gelechiid caterpillars, *Filatima sperryi* and *Gelechia mandella,* were doing a great deal of damage to the young foliage of plants in Valley County. After hatching, the first-instar caterpillars bore into the fruit and feed on the nutritious developing seeds, plugging the entry hole with their silk. Later instars bore out through a new hole and spin a silken web and tunnel to which they retreat when not feeding on leaves. The partially eaten leaves dry and cling to the silk, disguising and protecting the webs, which give the shrubs a silvery halo when backlighted.

The walnut spanworm (*Phigalia plumogeraria* [Hulst]) was discovered defoliating antelope bitterbrush plants in a seed orchard near Nephi, Utah, in June 1979.[27] In May the antelope bitterbrush plants appeared normal. Flower production had been excellent and a good seed crop was expected. By June 6 the plants in the orchard were completely defoliated. The spanworm larva is the usual geometrid type with five instars. The first-instar larvae are very small, only 1/16 of an inch long. They appear blackish, but under a hand lens the body can be seen to be mottled dark brown with a broken white lateral line. With each instar the length increases until by the fifth instar the larvae are 1 1/2 inches long and grayish or grayish yellow. The rather stout pupa is shiny dark brown and varies in length from 1/2 to 3/4 inch. Furniss first observed walnut spanworm eggs in the Utah seed orchard in April while the antelope bitterbrush plants were still dormant. Eggs were laid on the previous year's terminal shoots, mainly in the upper crown, in close-spaced masses of 23–345 eggs. Hatching began in the field in early May and was virtually complete by midmonth. The first-instar larvae dropped readily on silk threads when they encountered the end of a twig or leaf. This behavior aided in wind dispersal. Older larvae did not exhibit this behavior. Dispersal is important in this species because the adult females do not fly.

Friable soils are well suited for walnut spanworm larvae to enter and pupate, and large populations may build up in such soils; compact soils are less suitable. Perhaps most important, according to Furniss, is that vast populations of eggs or young caterpillars can be present, even in a small seed orchard, without being visible. Damage caused by the first two or three instars is subtle, usually consisting of etching on the leaves. The increase in physical size of the insect in the fourth and fifth instars suddenly makes these larvae very destructive herbivores. Antelope bitterbrush plants concentrated in relatively dense plantings can harbor severe outbreaks of this insect pest. Before the outbreak in the seed orchard at Nephi, entomologists did not even know that the walnut spanworm fed on antelope bitterbrush leaves.

The mountain mahogany looper (*Anacamptodes animata profanata* [Guen.]) periodically attains epidemic numbers on antelope bitterbrush.[28] Before 1961 this species was known only from adult moths collected in flight; its hosts and biology were unknown. Then it suddenly defoliated thousands of acres of curlleaf mountain mahogany (*Cercocarpus ledifolius*) along with intermingled antelope bitterbrush and other shrubs.[29] The discovery by a local rancher that mountain mahogany looper larvae were defoliating acres of browse species came 105 years after the adult moths were first collected and described. As for the outbreak, Furniss suggested that normally, as in the case of many insect herbivores, natural diseases and parasitic insects keep the populations of the mountain mahogany looper suppressed. For some reason, in the mid-1960s in the Juniper Mountain area of southwestern Idaho, the natural controls on population size did not function and the looper populations exploded. The shrubs were defoliated for at least three consecutive years. Many of the rose family shrubs were overmature when the infestation occurred. The additional stress of repeated defoliation weakened them and left them susceptible to bark insects. Furniss reported that the majority of the browse species were dead by the time the insect populations had declined. The major factor in the decline of the walnut spanworm was starvation of the larvae because they outgrew the potential food supply. At the same time, the local mule deer population crashed because they exceeded the winter range's carrying capacity.

The western tussock moth (*Orgyia vetusta gulosa* [Boisd.]), a very destructive pest in coniferous forests of the Pacific Northwest, has also been observed to defoliate antelope bitterbrush.[30] Western tussock moth infestations near Reno during 1958–1959 and again in 1963 and 1964 resulted in extensive antelope bitterbrush defoliation and twig death.[31]

During the late 1960s it was apparent that thousands of acres of antelope bitterbrush in northeastern California and adjacent Oregon were dying without significant stand renewal. With the benefit of hindsight it becomes obvious that much of this die-off was the result of old age, prolonged overutilization, increased competition from western juniper and cheatgrass, and general lack of ecological conditions conducive to stand renewal. As the antelope bitterbrush became old and senescent, increasing outbreaks of the eriophyid mite *Aceria kraftella* K. caused lateral branches to die back.[32] Sucking insects such as the bitterbrush tortoise scale (*Lecanium cersifex* Fitch) also caused branch dieback, sometimes resulting in partial or complete death of the host plant. Sucking insects serve as vectors for plant diseases as well. The bitterbrush tortoise scale was described in the mid-1970s as inhabiting Oregon, Idaho, and Montana,[33] but in 1996 we examined many antelope bitterbrush shrubs in northeastern California that were infested with these scales.

Ralph Holmgren reported that grasshoppers caused significant damage to antelope bitterbrush in the Payette and Boise River drainages of Idaho.[34] He did not specify which species of grasshoppers were involved.

In northeastern California, Hubbard reported a variegated cutworm (*Lycophotia margaritosa*) to have severely damaged antelope bitterbrush seedlings. In 1956 at Doyle, California, 90 percent of the emerging seedlings were lost, and 90 percent of this loss was attributed to cutworms and wireworms.[35] Cutworms are 1½–2 inches long and variable in color, but usually gray or brown mottled above with oblique gray areas on the sides. The adult moths are grayish brown with dark mottled forewings and a brassy luster. Wireworms, the larvae of click beetles, are long, slim, and cylindrical.

The diversity of insects found in antelope bitterbrush stands is a stabilizing factor in the plant communities.[36] Outbreaks of specific insects that can be damaging to browse species result from factors that upset the natural balance of the species. Furniss suggested that it is very important to understand the predators (including spiders) and parasites of the insects found on the leaves, flowers, and roots of antelope and desert bitterbrush and cliffrose. We have barely mentioned desert bitterbrush and cliffrose in this chapter because there is virtually nothing in the literature regarding the insect relations of these species.

Very little is known as well about the plant diseases that attack *Purshia* species. At the time of the big dieback of antelope bitterbrush in northeastern California, considerable effort was made to identify pathogens that might be the cause.[37] Numerous fungi were cultured from cankers associated with antelope bitterbrush stem dieback, and six were identified as pathogens.[38] Many of these pathogens may have entered already injured branches, however.

Perhaps the most studied of the diseases that affect antelope bitterbrush is the complex phenomenon known as damping-off, a disease of seedlings.[39] Infection usually begins just below the soil level on the upper portion of the taproot or the lower hypocotyl. Seedlings are most susceptible during rapid early growth when plant tissue is primarily parenchymatous, prior to the formation of lignified or suberized cell walls and secondary fibrovascular cambial activity. The pathogen kills the cells it invades, the root or hypocotyl tissue collapses, and the seedling falls over near the soil line. The hypocotyl appears water soaked or rotted and shrunken with a darkened discoloration. In general, cool and moist conditions are more conducive to damping-off induced by pythiaceous fungi, and slightly warmer and drier conditions support *Rhizoctonia*- and *Fusarium*-related damping-off or root rot.

Several researchers have commented on damping-off in antelope bitterbrush. R. A. Peterson was probably the first to report what he called "severe seed spoilage" of seedlings.[40] R. C. Holmgren, and R. C. Brown and C. F. Martinsen, reported damping-off symptoms in field trials in 1954 and 1959, respectively.[41] Nancy Shaw collected seedlings with apparent damping-off symptoms at the Forest Service Lucky Peak Nursery in Idaho, and David Nelson isolated *Fusarium, Rhizoctonia,* and especially *Pythium* from the material.[42] In his monograph, Eamor Nord reported that under controlled conditions *Rhizoctonia* and *Pythium* caused high mortality in inoculated antelope bitterbrush seedlings.

As nurseries began to grow container and bare root antelope bitterbrush seedlings for transplanting in the field, severe losses from apparent damping-off were reported. Some nurserymen thought they saw a relation between the seed source and susceptibility to damping-off. This suggested to Nelson that certain natural populations might be resistant to the pathogen. In screening seed sources from seven western states he isolated species of *Sphaeropsis* and *Sclerotium* that potentially could produce damping-off symptoms.[43] The importance of these two pathogens is they are transmitted *on* seeds.

Nelson artificially inoculated antelope bitterbrush seeds with *Sphaeropsis* and *Sclerotium* isolates from eight seed sources and compared the influence of the pathogens on germination and seedling survival.[44] One source was from Janesville, California, where the Lassen cultivar of antelope bitterbrush was first collected. None of the inoculated seedlings survived. Seeds from the Janesville plants showed significant reduction in germination with all of the isolates of both organisms. The pathogens significantly reduced germination in the other sources as well, but none was as uniformly reduced as the Janesville collection.

The fungus *Fomes annosus* (Fr.) Cke. causes death or heart-rot in virtually every tree species found in western forests.[45] In the eastern United States this disease is characteristic of conifer plantations planted on abandoned croplands. In the West, *Fomes* infections often occur in the stumps of conifers cut in logging operations. *Fomes* is also found in natural conifer communities where wind breakage or snow slides have damaged trees. Sporophores of the fungus have been found attached to the stems of antelope bitterbrush. It is not known how extensive or serious *Fomes* infection may be in natural *Purshia* stands, but it is a disease that should be considered in topping operations designed to increase or make available browse production. The cutting activity may spread the disease or make the shrubs more susceptible to infection.

Seedbed pathogens are similar to insect communities in that the presence of a diversity of organisms tends to prevent the dominance of any single organism. It is obvious, however, that we still have much to learn about plant pathogens and the ecology of *Purshia* species.

Chapter Ten

Wildfire Relations

One of the most influential factors in bitterbrush population dynamics is fire. Rangelands dominated by woody vegetation suggest the absence of fire; domination by herbaceous vegetation suggests the presence of fire. Herbaceous vegetation apparently dominated the landscape before Europeans arrived in the American West. Wildfires are a natural feature of the environment. During the past century and a half, human interference with fire and other natural processes has greatly influenced western vegetation.

As you proceed north and south from the equator, and until the timber line is reached with increasing latitude or altitude, woody plant biomass accumulations tend to increase over time in the form of standing wood and litter. The efficiency of production of woody biomass declines with altitude and latitude, but the rates of microbial decomposition and mineralization also decline. In tropical rain forests, litter decomposes almost as soon as it reaches the forest floor. Higher latitude forest, woodland, and shrub communities are too cold or dry for decomposition for significant portions of the year. The accumulation of woody biomass eventually leads to catastrophic recycling through burning in wildfires. Plant communities always reflect in their species composition, spatial distribution, and demography how the previous community that occupied the site was destroyed, which is known as the *stand renewal process.* If the stand renewal process has been consistently catastrophic over time, the natural plant community will have evolved under selection pressure associated with characteristics of the stand renewal process. Seed germination and seedling establishment will be adapted to the stand renewal process or a vegetative process; stem, root crown, or root sprouting will have been selected to renew the stand.

During most of the twentieth century there was a concentrated effort in the United States to suppress all wildfires on range and forest lands. Promiscuous burning was associated with the mismanagement of natural resources. Only in the last quarter of the century was it widely accepted that fire is a necessary part of wildland ecology. Thus, within little more than a century, wildland communities have known the natural frequency of wildfires, promiscuous burning, and complete suppression of wildfires.

In the past couple of decades it has become generally accepted that climates change on a long-, middle-, and short-term basis. Over the last two centuries, and perhaps for longer than that, human activities have been a factor in climatic change as well. It is of little wonder, then, that the wildfire relations of *Purshia* species are very difficult to interpret.

Native Americans are reported to have used fire as a tool for many purposes (e.g., communication, hunting, and forage manipulation). Certainly they knew that burned areas would soon grow grass and shrubs providing good forage for deer, elk, and, later, horses, and their use of fire for this purpose may have played an important role in forage succession. They may even have recognized the importance of bitterbrush to deer and elk and made efforts to establish or reestablish it.[1]

European settlers brought domestic livestock, which, in using restricted areas of range year after year, especially in the early spring, damaged perennial grasses and reduced the acreage these grasses covered. It was these herbaceous species that provided the fine fuels needed to carry fires. The reduction in native perennial grasses in turn reduced the competition seedlings of shrub species faced and increased their chances of establishment. The decrease in fire frequencies in the late nineteenth century, that is, resulted in the transition from herbaceous-dominated communities to landscapes dominated by woody species at the turn of the century.

The native perennial grasses that fueled wildfires did not dry sufficiently to carry fires until late summer and early fall.[2] Plants that flowered in the spring and matured in early summer, such as antelope bitterbrush, had a decided advantage over those, such as sagebrush, that flowered and set seed later in the year. Antelope bitterbrush seeds were safely cached in the soil before the wildfire season began, but late-season fires frequently destroyed the nonsprouting big sagebrush stands before seed was produced. Big sagebrush does not develop persistent seed banks, and the achenes have a deciduous pappus so they have limited wind dispersal and no reported granivore dispersal. Obviously, then, catastrophic wildfires hurt big sagebrush stands, and the species is not inherently equipped for rapid recolonization of burns.[3] This does not mean that repeatedly burned big sagebrush/bunchgrass communities were open to colonization by *Purshia* species. Antelope bitterbrush seedlings would have had a very difficult time competing in dense, established perennial grass stands.

Fire plays a crucial role in the seed and seedbed ecology of antelope bitterbrush. The timing of wildfires is of particular importance, as are the conditions of the year the fire occurs. In some years, antelope bitterbrush seed production may be low or nonexistent because of late frost, insect outbreaks, or excessive browsing the previous season. Because the seed banks of antelope bitterbrush are controlled by granivores, there may be no seed reserves available if the stand is burned in a poor seed production year. Likewise, if the stand burns while

Fig. 10.1. Bluebunch wheatgrass (*Pseudoroegneria spicata*) on this ridge top did not burn in a wildfire fueled by dry cheatgrass (*Bromus tectorum*) on the lower slopes. Invasion by the exotic cheatgrass has changed the chance of ignition, rate of spread, and season of wildfires.

the seed crop is still on the plants, all is lost. One week later, however, and the seeds may all be safely cached in the soil. Caches made in the litter halo beneath antelope bitterbrush crowns may be destroyed by wildfires, but caches in the spaces among shrubs are usually safe. Wildfires may control the distribution of sites suitable for the caching of antelope bitterbrush seeds by rodents (see Chapter 7). We will revisit fire-rodent–antelope bitterbrush relations later in this chapter.

Because the timing of wildfires is critical in the seed and seedbed dynamics of antelope bitterbrush, anything that affects wildfire timing also affects the plant's survival. The introduction of cheatgrass (*Bromus tectorum*) advanced the likelihood of wildfires from late to early summer. The fine-textured herbage of cheatgrass dries six weeks to two months before the native perennial grasses, and in years when it is abundant, the chances of ignition and rapid spread of wildfires are greatly enhanced (Fig. 10.1). This is disastrous for the seedbed dynamics of antelope bitterbrush.

Among the woody species that become transitory dominants of big sagebrush communities after burning are crown- or root-sprouting species of rabbitbrush (*Chrysothamnus*), horsebrush (*Tetradymia*), desert peach (*Prunus*), and ephedra (*Ephedra*). Rabbitbrush and horsebrush initially reestablish after wildfires by sprouting, but they dramatically increase in density through subsequent seed production and seedling establishment.[4] It is important to recognize that these semiwoody to woody transitory dominants of burned big sagebrush communities accomplish dominance through seed production from *first*-year sprouts. Antelope bitterbrush sprouts require years of growth before they flower.

As we indicated in Chapter 3, antelope bitterbrush is primarily a species of woodlands and their margins. It rarely ventures out into the true aridity of sagebrush-dominated communities. The postwildfire succession of *Purshia-Artemisia* communities is vital to the management of antelope bitterbrush for two basic reasons. First, woodland communities on the border of sagebrush environments are *initially* sagebrush communities until the trees overtop the shrubs. Second, the essential antelope bitterbrush communities for mule deer are the winter browse communities that are available during very severe winters. By their physical location, these essential communities are usually sagebrush–antelope bitterbrush communities.

August Hormay, in his pioneering research on antelope bitterbrush, had a very simplistic view regarding the influence of wildfires: "Fire kills bitterbrush in most instances in California."[5] Although he agreed, Eamor Nord took a much more cautious approach in his monograph on bitterbrush.[6] He cited references to research by such noted range researchers of the time as Don Cornelius, Joe Pechanec, and James Blaisdell to support the concept that burning was bad for antelope bitterbrush. He also cited a recent publication by Blaisdell and Mueggler, however, indicating that sprouting of antelope bitterbrush was common on the eastern Snake River Plains following fire.[7]

Perhaps to determine which of this conflicting information on bitterbrush sprouting was correct, Nord looked at desert and antelope bitterbrush communities in California and Nevada that had burned during the previous 32 years. Only 1 of the 11 sites Nord studied had completely recovered after burning, and that burn was 32 years old. Nord also observed that burns had to be at least 10 years old before antelope bitterbrush reestablished and produced seed. Immediately after burns most plants resprouted, but postburn seedling establishment was highly variable. Seedlings constituted from 5 to 80 percent of the total plants present in the first 5 years after burning. Nord also noted that wildlife was attracted to the sprouting vegetation in burns.

More is known about the fire history of ponderosa pine woodlands than any other environment where antelope bitterbrush occurs. Ponderosa pine trees are readily scarred by wildfires, and thus provide a record of the area's fire history.

Fires that damage a part of the basal trunk cambium but do not destroy or kill the tree leave a scar, or cat-face, on the tree. Some ponderosa pines have been repeatedly scarred by wildfires. Each time the trunk is damaged, the adjacent undamaged cambium tries to grow over the burned area, leaving a characteristic mark in the growth rings that makes it possible to date repeated wildfires. Such scars are not perfect chronologies of wildfires. Fire-scarred trees usually occur in areas that are somewhat protected from fire; trees in very fire susceptible situations are generally killed by the first fire. Chronologies from single trees may be combined with those from other fire-scarred trees from a wider geographic area to construct regional fire chronologies. And while they are not perfect, such chronologies are much better than any system that has been derived for treeless environments such as big sagebrush–antelope bitterbrush communities.

When Harold Weaver's classic paper "Fire as an Ecological and Silvicultural Factor in the Ponderosa Pine Region of the Pacific Slope" was first published in 1943, it was considered heresy by many professional foresters.[8] Fire was the enemy in those days, an evil to be stopped at all costs. Most of the early forestry bulletins published about far western pine woodlands include a photograph of a fire-scarred ponderosa pine with a caption bemoaning the terrible waste of timber resources these scars represent.

Although Weaver believed that shrub understories, such as antelope bitterbrush, had been present in the ponderosa pine woodlands of the 1880s, he thought that frequent ground fires (every 11–47 years) kept them minor components and maintained a parklike appearance in the woodlands. Precontact ponderosa pine woodlands were, he said, "a place where you could easily drive a wagon." In the first half of the twentieth century extremely thick patches of pine reproduction were sometimes encountered in ponderosa pine woodlands. Called "dog hair patches" by loggers and ranchers, they were attributed to the lack of ground fires in the woodlands. Robert Rummell disagreed. Rummell studied an ungrazed woodland on Meek's Table, a mesa in central Washington, with an understory dominated by pinegrass (*Calamagrostis rubescens*) and elk sedge (*Carex geyeri*). He attributed the abundant pine reproduction to grazing rather than a lack of wildfires, although the mesa had not been burned over in the previous 125 years.[9]

Not everyone agreed with Weaver's assessment of wildfire frequency. S. F. Arno, working in the Bitterroot National Forest of Montana, arrived at a much shorter frequency of wildfires—every 4–11 years.[10] C. H. Driver et al. considered the frequency of wildfires in the precontact ponderosa pine–antelope bitterbrush woodlands of Washington to be in the 7–10-year range.[11]

And A. H. Johnson and G. A. Smothers estimated the wildfire interval in the ponderosa pine forest of the Lavabeds National Monument of northeastern California to be as low as 6 years.[12]

Because the art and science of interpreting precontact wildfire intervals from fire-scarred tree sections is not precise, Henry Wright suggested that the basic ecology of the species involved should be considered in interpreting such chronologies.[13] Antelope bitterbrush, for example, must be at least 10 years old before it flowers, and even then seed production is very limited. Curlleaf mountain mahogany requires 30–50 years before it flowers. In big sagebrush communities the crown- or root-sprouting species of rabbitbrush or horsebrush are the woody dominants for a transition period of 10–15 years after wildfires. If the sagebrush wildfire interval had been shorter than this, there would have been no sagebrush communities when Europeans arrived. The hardest part for the precontact wildfire chronologies to interpret is the spatial relations of the fires.

Changes in the distribution and tree density of western juniper and pinyon-juniper woodlands have had a dramatic influence on *Purshia* communities. Tom Johnson's classic paper "One-Seeded Juniper Invasion of Northern Arizona" offers a very helpful explanation for what happened to pinyon-juniper woodlands during the twentieth century.[14] During the late-nineteenth-century mining booms, pinyon and juniper trees were harvested and converted to charcoal for use in the reduction of silver ores. Vast acreages of pinyon and juniper were removed around large mining centers such as Eureka, Nevada,[15] probably associated with widespread promiscuous burning to dispose of slash. The removal of the trees created ideal conditions for shrub succession.

Juniper Hill, in northwestern Lassen County, California, was a superb mule deer wintering area during the first half of the twentieth century. The region consisted of a low range of hills at the margin of Big Valley with vegetation making a transition from ponderosa and Jeffrey pine woodlands in the Eagle Lake uplands to the agricultural valley below. The hills are the typical basalt of the Modoc Plateau. Most of the slopes are gentle, but rimrock occurs along the higher ridges. In the 1970s we initiated a study of the western juniper woodlands on Juniper Hill that included demography studies of the juniper woodlands and construction of precontact wildfire chronologies. The publications that resulted from these studies have been very unpopular with the scientists and natural resource managers who strongly support prescribed burning as a cure-all for the ills of mule deer habitats.

Juniper Hill once provided near ideal winter range for mule deer in terms of the amount and quality of browse production (antelope bitterbrush and mountain big sagebrush) and geographic location below the normal deep snow

Fig. 10.2.a. The Parks family ranch at Juniper Hill, Lassen County, California, ca. 1900. Note scattered western juniper trees on the ridge in the background. Other vegetation on the ridge is mountain big sagebrush and antelope bitterbrush.

line (Fig. 10.2.a). In the 1920s the townspeople of Adin, California, would be disappointed if they did not see hundreds of mule deer (some said thousands) on a winter afternoon's drive around Juniper Hill. The antelope bitterbrush stands on Juniper Hill were so tall and extensive that the children of the Parks family, who lived on the east side of the hills, had a hard time finding the family milk cows.

In the 1970s, Juniper Hill supported extensive stands of western juniper trees approaching 30–35 feet tall with about 40 percent crown cover (Fig. 10.2.b). Beneath and between the juniper trees were numerous skeletons of antelope bitterbrush and mountain big sagebrush plants. Root studies revealed that although the crown cover of the juniper trees was only 40 percent, the soil profile was totally occupied by tree roots both horizontally and vertically. Soil and vegetation studies revealed two plant communities on Juniper Hill. About 70 percent of the area had relatively deep residual soils with loam-textured surface soils. It was these sites that had once supported the very productive antelope bitterbrush–mountain big sagebrush communities. By the 1970s these sites

Fig. 10.2.b. The same site photographed in 1978. The foundations of the house are located behind the first tree across the fence.

had about 150 maturing western juniper trees per acre. The remaining 30 percent of the hill supported low sagebrush communities with about 28 western juniper trees per acre growing on shallower soils with clay-textured surface soils. It has been suggested that the low sagebrush sites are former big sagebrush sites whose surface soil horizon was removed by geologic erosion.

The big sagebrush–low sagebrush dichotomy is a common feature in western juniper woodlands and adjacent shrub communities (see Chapter 3). In the 1970s, 84 percent of the juniper trees present in the big sagebrush potential communities on Juniper Hill had become established between 1890 and 1920 (Fig. 10.3). The oldest tree was established in 1855, and only 6.4 percent had established before 1890. Only 2.8 percent of the existing trees had established since 1930. We found no naturally dead trees in these communities. The western juniper trees growing in the low sagebrush communities had a much greater average age. The oldest trees, about 2 percent of the population, were established by 1600, and most were established before 1800. Only 17 percent of the trees in the low sagebrush had become established since 1900, compared with 98 percent in the big sagebrush communities. Obviously, these demography

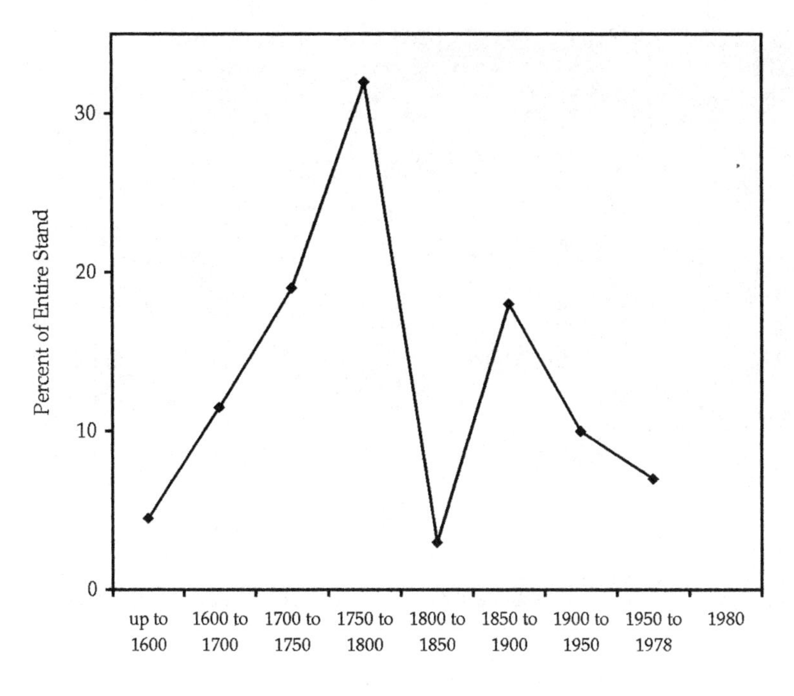

Fig. 10.3. Periodicity of establishment of western juniper trees in low sagebrush communities on Juniper Hill. Adapted from James A. Young and Raymond A. Evans, *Germination of Seeds of Antelope Bitterbrush, Desert Bitterbrush, and Cliffrose* (Agric. Res. Results 17, USDA, Agric. Res. Ser., Oakland, Calif., 1981).

data show that the conversion from big sagebrush–antelope bitterbrush on the big sagebrush sites occurred during the twentieth century.

Several studies have shown that western juniper trees up to 50 years old are very susceptible to wildfires.[16] Our sample of 28 fire scars from Juniper Hill revealed three decades (1640–1650, 1750–1760, and 1830–1840) when more than two trees were fire scarred (Fig. 10.4.a). Each of these instances of wildfire probably represents one large fire. The variation in dates within decades from which scars are identified is probably due to false or missing rings.

We found one remarkable tree that was doubly scarred by a fire between 1750 and 1760. The cambium was completely destroyed on opposite sides of the tree. Essentially, the tree trunks were separate trunks at the base. On these twin trunks were separate, but synchronous, fire scars for the decades of 1770, 1780, 1790, 1830, and 1850. Although this tree was in a topoedaphic situation where it was susceptible to fire scarring in five successive fires, it grew from 1655 to

Fig. 10.4.a. Frequency of fire scars on western juniper trees growing on low sagebrush sites on Juniper Hill. Adapted from Young and Evans, *Germination of Seeds*.

Fig. 10.4.b. Cross section of a fire-scarred western juniper trunk. Such scars record repeated wildfires.

1778 without evidence of such scars. If fire alone controlled the distribution of western juniper on the big sagebrush sites, the century without fire should have allowed the big sagebrush potential sites to be occupied by woodlands. It seems probable that at least one of these trees would have survived subsequent wildfires. But the only old-growth western junipers growing on big sagebrush potential communities were located in rimrock areas with low sagebrush communities downslope. These stands contained ax-cut stumps of trees cut by homesteaders to be split into posts for fencing. Some of these stumps had been cut above limbs that then assumed apical dominance and subsequently grew upright, allowing us to roughly date the time of cutting. These sites were apparently safe from fire because of their topoedaphic situation, but 85 percent of the stumps had evidence of two fire scars that were roughly dated to the 1750 and 1830 fires.

Almost all of the fire scars we analyzed came from low sagebrush communities, but only 4 percent of the trees in the low sagebrush communities were fire scarred. The old trees in the low sagebrush communities were not cut for posts because the centers of the trees were infested with brown cubical rot. During the twentieth century, the low sagebrush sites have been considered wildfire proof because they lack sufficient herbaceous vegetation to carry fires between the shrubs and trees. Obviously, before Europeans arrived there was enough herbaceous fuel to allow occasional wildfires to scar the infrequent western juniper trees.

We observed in the Juniper Hill area excellent stands of crested wheatgrass (*Agropyron desertorum*) and sod-bound intermediate wheatgrass (*Elytriga intermedia*) that were being invaded by seedlings of western juniper. Herbaceous competition does not appear to be a major factor inhibiting western juniper establishment in this environment. Obviously, excessive grazing of the perennial grasses that would lead to their disappearance from the communities would reduce the chance of the ignition and spread of wildfires. Once western juniper trees dominate big sagebrush sites and have purged the understory of shrub and herbaceous vegetation, the stands become fireproof except for fires that can spread from tree crown to tree crown.[17] Remember, the "closed" stands of western juniper on the big sagebrush potential sites on Juniper Hill had only 40 percent crown cover.

There was no evidence that the pioneer ranchers whose homes were located on the flanks of Juniper Hill engaged in promiscuous burning. The location of the farmsteads would have made such burning hazardous to neighbors. You could conclude from all this that after settlement, wildfires were suppressed on the hill and western junipers increased to occupy the site totally. This may be a true synopsis of what happened, but it begs the question of why western juniper fruits (false berry) were available and suddenly dispersed rather evenly across this large acreage. A closer look at this question makes it obvious both that something besides wildfire frequency had to have changed in the general environmental setting of Juniper Hill, and that similar changes have occurred throughout the range of western juniper and perhaps throughout the range of pinyon-juniper woodlands.

Conifers are ancient plants. There are many monospecific genera of conifers, often the last remnant of their lines. Junipers evolved in the midlatitude, temperate desert environments that developed at the end of the Pleistocene and exhibit characteristics that reflect those conditions. Leaf scales, evolved from needles, improve moisture relations, and a fleshy covering protects the seeds. At maturity this false fruit is attractive to birds and mammals, which disperse the seeds over the landscape. Old-timers in the Juniper Hill area claim that western juniper seedlings were rare before the winter of 1889–1890, an exceptionally severe winter with huge amounts of snow and prolonged Arctic cold. As ranchers struggled to ration out hay supplies to keep livestock alive, they noted, for the first time, thousands of robins (*Turdus migratorius*) feeding on an abundant crop of western juniper berries. Over the next few years ranchers noted abundant juniper seedlings wherever the robins had roosted. The fields were fenced with split rail fences erected in zigzag patterns, and in many areas around Juniper Hill you can see western juniper trees lining fields in the zigzag pattern even though the old fences are gone. Perhaps this one exceptionally severe winter was enough to establish the robin–western juniper seed interaction, or perhaps some unrelated human or natural event set in place a dynamic increase in robin populations that in turn resulted in the interaction with the tree seeds.

Dynamic increases in western juniper seedling establishment require seed dispersal, and longer distance seed dispersal requires animal vectors and an abundance of seeds. Juniper seed production is highly erratic and varies among trees and from year to year. To support large populations of seed predators, there has to be exceptional production of western juniper seeds. The junipers may be evolving conifers in some aspects of their physiology, but they are still saddled with an erratic wind pollination system. The ultimate key in western juniper enhanced dispersal is what governs the initiation of viable male and female reproductive tissue on the trees. The rapid conversion of the sagebrush–antelope bitterbrush communities to western juniper woodlands must have been a function of animal-aided dispersal of the seeds. We found no evidence on Juniper Hill of isolated western juniper trees becoming established, reaching sexual maturity, shedding seeds, and being surrounded by progeny, and we are fairly certain that the juniper woodlands did not become established in this fashion. The stand demography indicates that the trees established in waves of nearly evenly aged seedlings.

We know from the recent geological record that western juniper has expanded and contracted its range several times during the Holocene.[18] The precontact changes in range are attributed to climatic changes. Perhaps human activity in the form of grazing livestock and wildfire suppression interacted with a subtle climatic change to produce the increase in western juniper and the corresponding reduction in antelope bitterbrush on Juniper Hill. Human-induced changes in atmospheric gases during the late nineteenth and twentieth centuries are well documented. Perhaps western juniper trees are responding to increased carbon dioxide concentrations in the atmosphere.

The explosive growth of the western juniper stand in the twentieth century begs another question: Why did the antelope bitterbrush population on Juniper Hill, which the junipers replaced, explode in the late nineteenth century? Perhaps many of the same factors that governed the ultimate increase in western juniper also conditioned the prior increase in antelope bitterbrush. Suppression of wildfire, reduction in competition from perennial grasses, and reduction in predation from mule deer populations reduced by excessive hunting would all have favored antelope bitterbrush. The bitterbrush increased before the western juniper because it reaches sexual maturity about twice as fast and, as a dicot, has a much more efficient—or at least consistent—reproductive system. As grazing pressure increased on Juniper Hill and the native herbaceous vegetation largely disappeared, the western juniper seedlings and young plants were protected from browsing because the juvenile vegetation has sharp needles rather than the mature plant's leaf scales. Perhaps precontact populations of mule deer were extremely small *because* recurring wildfires were a frequent part of the environment. If this had been the case, there should have been periods on Juniper Hill when wildfires failed to scar the trees, and shrubs and mule deer both increased.

We tried to test this theory by creating quarter-acre openings in the western juniper stands on Juniper Hill and attempting to establish herbaceous and browse species in these openings. Some wintering mule deer still find their way to Juniper Hill, but the dense woodlands also provide cover for a small year-round herd that largely feeds in neighboring alfalfa (*Medicago sativa*) fields. Although there were not many mule deer in residence on Juniper Hill, there was a sufficient number to completely destroy direct-seeded and transplanted browse species and even to influence the composition of the perennial grasses that established in the openings. We successfully transplanted an ecotype of fourwing saltbush (*Atriplex canescens*) that was not preferred by mule deer. Other ecotypes of this shrub that were preferred were lost to excessive browsing. The nonpreferred ecotype reached a height of about 3 feet before a severe winter with extended snow cover changed the relative preferences of the mule deer and the fourwing was heavily browsed and eventually killed. Never underestimate the potential of mule deer to change the species composition and dominance of herbaceous and woody vegetation![19]

What are the lessons of Juniper Hill? Obviously, western juniper–antelope bitterbrush succession is a tangled web that requires more study. Simplistic answers such as universal burning and wildfire suppression cannot stand up to field tests. It should be remembered that on Juniper Hill, the fire-scarred trees were almost entirely in low sagebrush environments, but the productive antelope bitterbrush communities were on contrasting soil environments where they shared dominance with big sagebrush.

Woody species that evolve in environments that are repeatedly burned are selected for characteristics that are adapted to burning; the woody lignotubers of manzanita (*Arctostaphylos*) are an example.[20] The extreme density of the wood of these tubers makes them virtually fireproof. The tubers store energy and buds that produce sprouts after the stand is burned. Manzanita and *Ceanothus* both produce seeds with dormancy mechanisms that are overcome by fire. We mentioned above that rabbitbrush and horsebrush plants combine sprouting with superabundant seed production to colonize postwildfire sagebrush sites. What, if any, adaptations to wildfires are apparent in *Purshia* populations?

Some woody species of chaparral communities are predetermined by their physical and chemical nature to burn. Species of *Adenostoma,* a rose family shrub that grows in California, contain oils that distil during burning with explosive results. Combining abundant sprouting with this explosive burning, the *Adenostoma* species are truly adapted to stand renewal by burning. Carol Rice, who examined antelope bitterbrush as fuel, outlined five points to consider when evaluating the fuel characteristics of bitterbrush communities:[21]

1. Fuel loading
2. Arrangement of fuel (horizontal and vertical continuity)
3. Distribution of size classes of fuel
4. Moisture content of fuel
5. Heat and ash content of fuel and volatile oil content

According to Rice, the available literature indicates that fuel loading in antelope bitterbrush communities is generally light. The fuel loading in antelope bitterbrush–sagebrush and antelope bitterbrush–western juniper types is so low that fire will not carry in these communities except under exceptional conditions. A. H. Johnson and G. A. Smothers suggested, in contrast, that the presence of abundant antelope bitterbrush in the understory of pine woodlands in Lavabeds National Monument increased the available fuel load to such a high level that a conflagration could destroy the entire community.[22] Martin reported that fuel loads ranged from 1.4 to 4.3 tons per acre in antelope bitterbrush–western juniper communities in the Pacific Northwest.[23] He considered these fuel loads to be very light and to consist mostly of perennial bunchgrasses.

Fuel load is a measure of how the components of a plant community influence fire characteristics and the effects of fire. In most antelope bitterbrush–sagebrush and open western juniper stands that still contain living antelope bitterbrush plants, the main trunk and primary branches of the bitterbrush plants are not consumed in fires. Apparently, the fire does not remain in one spot long enough to ignite the very dense old-growth wood. In terms of seedbed ecology, it is the fire characteristics of the litter beneath the shrub canopies that is important. Burning of these litter accumulations influences the physical, chemical, and biological characteristics of the postburn seedbed.[24] Mature antelope bitterbrush plants with erect growth forms accumulate more litter than individual big sagebrush plants.

Rice did not discuss the most important fuel load characteristic of antelope bitterbrush communities: cheatgrass. The sparseness of the woody fuel load is irrelevant if the understory is dominated by cheatgrass. In a year with exceptional growth of cheatgrass, a site will burn once ignited, and there is little that fire suppression agencies can do to prevent the loss of the shrubs. Furthermore, exceptional cheatgrass growth years carry over to the next season or seasons as dangerous fuel loads. Too many habitat managers have made the mistake of preventing domestic livestock from grazing the cheatgrass. This was the case in the Bass Hill wildfire that resulted in such exceptional flame heights and destructive consumption of the famous Lassen antelope bitterbrush stands. The southeastern end of Bass Hill was private rangeland at the time of the fire. The herbaceous and antelope bitterbrush stands on this private land were overutilized, with the antelope bitterbrush severely hedged, but after the disastrous wildfire on the refuge these adjacent, grazed stands were still intact. This does not mean that antelope bitterbrush–big sagebrush stands should be overgrazed, but it does mean that cheatgrass accumulations in antelope bitterbrush communities should be actively managed.

The arrangement of fuels is highly variable in *Purshia* communities because of the wide variety of growth forms the plants can assume. The decumbent types have a spreading crown with many fine limbs located close to the soil surface and to fine-textured herbaceous fuels. Upright forms in ponderosa pine woodlands can accumulate needles in their canopies that flare up, igniting the tree overstory.[25] The decumbent forms of antelope bitterbrush that have air-layering stems also have a specific distribution of fuel loads and subsequent reaction to burning. Dick Driscoll found that air-layering plants that sprouted after a wildfire always sprouted at the point of air layering and never at the main stem where the maximum accumulation of litter occurred.[26] Robert Clark and

Carlton Britton noted that adventitious buds of the decumbent forms had a different location around the trunk base than those of erect plants.[27]

Rice found the preponderance of "dead fuel" in antelope bitterbrush stands to be mostly less than ½ inch in diameter. We are not sure, however, what she meant by dead fuel. It could be literally dead material, or she could be referring to heartwood versus sapwood. We strongly disagree with either assumption. The main trunks of erect forms of antelope bitterbrush commonly reach 4–6 inches in diameter, with the main lateral branches being 1–2 inches in diameter. The main stems often do not ignite, but if they are totally consumed, they burn for a considerable period. Citing Green, however, Rice described antelope bitterbrush as a fuel that generally maintained the fire for less than an hour.[28]

Robert Martin et al. reported the average biomass of four northwestern shrubs by fuel size class and crown cover.[29] Antelope bitterbrush had the lowest average fuel load, and greenleaf manzanita the highest. This research note includes a table that estimates fuel load in tons per acre by percentage crown cover and provides lag time for the burning of the fuel classes. The tremendous variability in the morphology of antelope bitterbrush plants and the proportion of living and dead wood within given growth types as a result of age and condition make generalizations about the diameter classes of fuel loading mere generalized speculation. Anyone considering prescribed burning or fire suppression in antelope bitterbrush stands should consult a fire manager trained and experienced in the geographical area.

Nord and Countryman used ash and moisture content as indexes of the flammability of shrubs.[30] Compared with eight other species, the ash content of antelope bitterbrush was low. High ash content was correlated with resistance to burning.

C. W. Philpot determined the energy (heat) content of antelope bitterbrush to be average for shrubs.[31] The heat content averaged 5,040 calories per gram in the spring, declined to 4,600 calories per gram in summer, and then rose to 5,010 calories per gram in the fall.

Prescribed burns in antelope bitterbrush stands on the Fort Rock Ranger District in Oregon produced variable results.[32] The off–peak fire season burns were conducted in April or September. The young stands did not carry fires, and the stands with plants more than 40 years old were completely consumed by the fires.

It would appear that cliffrose and desert bitterbrush are more adapted to wildfire as a stand renewal process than antelope bitterbrush because they consistently sprout once the aerial portion of the plant is removed. Remember that the outcrossing nature of *Purshia* populations ensures that many characteristics are segregating through populations. It is possible that sprouting after top removal is a genetically controlled characteristic, and the potential for expression of this characteristic may exist in some portions of antelope bitterbrush populations and be absent from others.

The contradictory opinions in the literature regarding *Purshia*'s response to fire seem to indicate considerable geographic variability. Dwight Billings, for example, said that wildfires had eliminated antelope bitterbrush from the western Great Basin.[33] Blaisdell and Mueggler's famous paper on the Upper Snake River Plains and a lesser known paper by Dick Driscoll on central Oregon both indicate that sprouting of antelope bitterbrush after fire is common and significant in some stands.[34]

Starker Leopold narrowed down the geographic range of variability in describing his experiences with fire and bitterbrush during the 1950s. Logging of ponderosa–Jeffrey pine woodlands and recurrent fires stimulated bitterbrush growth near the town of Boca, California, in the Truckee River Canyon. A few miles down the canyon toward more arid Reno, Nevada, however, fires seemed to wipe out both bitterbrush and sagebrush, giving rise to almost pure stands of cheatgrass.[35] Obviously, if we are to have any understanding of the interaction between fire and *Purshia,* we have to stratify communities with regard to their environmental and genetic potentials.

J. P. Blaisdell initiated the concept that sprouting of burned antelope bitterbrush plants following fire is related to the intensity of the fire.[36] Later, he conducted prescribed burning trials with Walter Mueggler in big sagebrush and three-tip sagebrush (*Artemisia tripartia*)–bluebunch communities on the Upper Snake River Plains.[37] In the early 1950s, when these studies were conducted, cheatgrass was not a major component of the area. The researchers cut or burned antelope bitterbrush plants both in a relatively undisturbed community and in one that had burned in 1939 where antelope bitterbrush had reestablished by sprouting. They repeated the treatments on six dates from May through October. The antelope bitterbrush plants were burned with a weed burner rather than in a prescribed burn. Blaisdell and Mueggler described two types of sprouting. Some plants sprouted from a mass of dormant buds that wholly or partially encircled the stem at ground level. The second source of buds came from the callus of meristematic tissue usually located about an inch above the ground surface. The buds or meristematic callus that produced sprouts often appeared to come from stems that were protected by unburned bark. Sprouts from meristematic callus were slower to develop than those from crown buds. Both types of sprouting occurred in both cut plants and burned plants, but crown bud sprouts were the most common form in cut plants.

The date of clipping had little effect on sprouting, but burned plants had the highest sprouting rates after the May and June burns. At every date there was more sprouting from the plants in the area previously burned in the 1939 fire.

A total of 90 bushes (75 percent) sprouted in the old burn and 56 (50 percent) sprouted in the area not previously burned. The subsequent survival of the antelope bitterbrush was fairly poor. Only 108 of the 146 plants that sprouted were alive the next year. Mortality was particularly high among plants that sprouted after burning in the May and June treatments.

The loss of antelope bitterbrush sprouts the second or third season following a fire has been a problem in many areas. Fred Wagstaff studied succession in an important mule deer wintering area in Wasatch County, Utah, four years after it had burned in a large fire in 1979.[38] There had been considerable sprouting, but in a burned-unburned comparison, the browse production in the burned area was practically nil because of the small size of the plants and their coverage by even a slight snowfall. Remember, too, that Nord stressed how attractive sprouting browse in burns is to animals and the damage that concentrations of animals can do to postburn succession.

In 1994 John Cook and his colleagues burned high-elevation (ca. 8,000 feet) mountain brush communities and studied the response of the vegetation.[39] This modern study has additional significance because it was conducted in a geographical area where little has been written about *Purshia* species. The antelope bitterbrush was entirely eliminated by the fire in some locations, while in others as many as 66 percent of the plants survived. It is not clear if the surviving plants were regenerated from sprouts or were plants that escaped the fires. Within three years the surviving plants had sufficiently increased browse production to compensate for the lost vegetation. The authors found that the crude protein content of herbaceous vegetation in the burned areas was markedly higher than in paired unburned controls. Increased protein content may be significant in attracting concentrations of grazers and browsers to the vegetation in burned areas and therefore influencing succession.

At another extreme of the distribution of antelope bitterbrush, W. W. Fraas et al. studied the effects of prescribed burning on an antelope bitterbrush–mountain big sagebrush/bluebunch wheatgrass community in the Steep Mountain area near Butte, Montana.[40] They sampled burned and unburned communities eight years after the prescribed burn. Antelope bitterbrush density did not differ between burned and unburned communities, but cover, flower production, and seed production were found to be less on the burned site.

James Blaisdell began his research on antelope bitterbrush during the early 1950s on the Upper Snake River Plains in response to complaints from wildlife managers that prescribed burning was destroying mule deer habitat. By the 1970s, prescribed burning was gaining acceptance among the managers of ponderosa pine woodlands in the Pacific Northwest. Robert E. Martin, then with the USDA, Forest Service, and Charles H. Driver, a professor of forestry at the University of Washington, were among the major advocates of prescribed burning in woodland management, but they were concerned that criticism from wildlife managers was limiting the application of prescribed burning. To allay such criticism they prepared a lengthy review of the factors affecting antelope bitterbrush reestablishment following fire.[41] The review was divided into sections on sprouting and seedling establishment. Under sprouting they discussed the following points:

1. Genetic variability and morphology
2. Phenological condition
3. Plant age
4. Competition
5. Soil type
6. Soil moisture
7. Burning conditions
8. History
9. Fuel load
10. Browsing pressure

The genetic variability section summarizes studies of the variability of *Purshia* populations, including Wagle's lengthy dissertation from the University of California at Berkeley.[42] But neither the cited studies nor subsequent work has established that sprouting is a heritable characteristic in *Purshia* populations. Without citation, Martin and Driver reported that sprouting of antelope bitterbrush can occur from 3 to 13 months after fire.

Martin and Driver also reported that antelope bitterbrush can develop lignotubers—swellings of the main stem at or just below the soil surface—that consistently sprout. They hedged their bet, however, by including a disclaimer that it was possible that lignotubers occurred only in central Washington, where Driver et al. had found them.[43]

Martin and Driver had a hard time with the phenology and sprouting sections. They reviewed the applicable literature indicating that total available carbohydrate reserves decrease with bud breaking, growth, and flowering (see Chapter 4). The problem is that some researchers (e.g., Blaisdell and Mueggler) have found that the best sprouting occurs with late spring burning when carbohydrate reserves would be at their lowest. Apparently, no plant physiologist has ever considered that growth regulators might control sprouting of *Purshia* species. Such growth regulators might be most abundant when the shrubs are actively growing.

Martin and Driver reported that very young and very old antelope bitterbrush plants (i.e., older than 5 years and younger than 40–60 years) sprout extensively. Citing Blaisdell and Mueggler, they further indicated that plants between 5 and 15 years of age are much more likely to sprout than older plants.

They also noted that antelope bitterbrush plants increase in browse production until they are 60–70 years old, although they included a record of a 92-year-old antelope bitterbrush sprouting when cut 6 inches above the soil surface.[44]

With regard to the effect of soil type on sprouting, Martin and Driver disagreed with Driscoll and concluded that fine-textured soils produce more sprouting because of the higher soil moisture content. They agreed with Driscoll, however, that the least sprouting occurs where pumice and cinder-sized volcanic tephra are found on the soil surface.

A number of researchers have found that postfire sprouting is more extensive in the spring and fall because the higher soil moisture in those seasons produces cooler fires.[45] Obviously, the conditions under which a fire, either wild or prescribed, burns an antelope bitterbrush stand will affect any subsequent sprouting. The problem Martin and Driver encountered in attempting to summarize the information was that most of the older articles on prescribed burning do not measure or otherwise quantify the fire. For example, the term *intensity* as used by Blaisdell is best equated to the modern term *rate of heat release per unit of fire front*. But this may not be related to the amount of heat delivered to the crown of the plant.

Martin and Driver's "History" category is rather vague. Basically, it follows Driver's assumptions that repeatedly burned communities should be more adapted to survival and renewing the preexisting community where fire is rare. Perhaps this smacks of Lysenkoism.

Considering the background of the authors, we expected a lengthy review of fuel loads in relation to bitterbrush sprouting. Instead, Martin referred to his paper indicating that there is sufficient fuel in pine woodlands so that even with prescribed burns conducted under cool, moist conditions, there may be little sprouting of antelope bitterbrush.[46]

In the final category, Martin and Driver went to considerable lengths to relate the preexisting level of utilization of antelope bitterbrush plants to the potential for sprouting after burning. That literature is reviewed at several other locations in this manuscript. The authors had no direct evidence to relate the level of utilization to sprouting potential.

Writing the review paper seems to have convinced Martin that sprouting is not a viable alternative for renewing antelope stands after prescribed burns. At the same symposium at which the review was presented, Martin presented a paper on successful restocking of prescribed burns through the establishment of seedling antelope bitterbrush.[47] In that paper Martin emphasized the caching of antelope bitterbrush seeds from unburned borders of prescribed burns as a successful method of restocking. And indeed, as we mentioned earlier in this chapter, timing burns so they occur *after* the current year's crop of seeds are mature, dispersed, and cached may be the most important factor in guaranteeing postwildfire succession in these communities. This applies to big sagebrush and western juniper–big sagebrush communities, but it is possible that the fuel in ponderosa pine woodlands might generate fires hot enough to kill the cached seeds. Because rodents do not cache seeds in heavy pine litter, however, some caches should survive even in pine woodlands.

In his study of the effects of prescribed burning on control of big sagebrush, Blaisdell found that bitterbrush production was less in two burned areas 12 and 15 years after the prescribed burns than it had been before the burns.[48] In 1979, Robert Murray sampled an Upper Snake River Plains area burned in 1936.[49] After 43 years, the production of antelope bitterbrush on a similar unburned area was nearly twice that recorded for the burned area. The very limited antelope bitterbrush seedling recruitment on the burned area Murray attributed to an increase in native perennial grasses that were released from sagebrush competition by the fire. In noting this, Murray hit on an aspect of fire and antelope bitterbrush ecology that is becoming increasingly apparent in the Intermountain area. For many years, repeated burning of big sagebrush communities coupled with continued heavy grazing encouraged cheatgrass dominance.[50] If enough perennial grasses remain in the stand to occupy the environmental potential released by the destruction of the shrubs, however, and if proper post-burn livestock management is practiced, perennial grasses will dominate the site for some transitory period.[51] Since the 1960s, grazing management systems for domestic livestock have been implemented for virtually all federally owned rangelands. The combination of reduced livestock numbers and post-wildfire management systems that usually require two years' rest from grazing has resulted in a tremendous increase of perennial grasses on many higher elevation ranges in the Intermountain area—in areas where there are enough perennial grasses left to restock the stands. These are generally the areas with greater effective moisture for plant growth (e.g., higher elevation and north-aspect sites). On lower elevation sites where the stands of native perennial grasses have been virtually eliminated and no seed banks exist, perennial grasses have not returned and the management systems have favored the continued dominance of cheatgrass. Some higher elevation sites have even reached the successional stage at which repeat burns are occurring in *perennial* grass stands, perpetuating perennial grass dominance. Neither cheatgrass stands nor perennial grass–dominated stands are conducive to the establishment of antelope bitterbrush seedlings. Both vegetation types are capable of providing fuels for wildfires that will destroy bitterbrush stands even though the fires will probably occur at different seasons.

Eric Loft and his coworkers captured many wildlife managers' fears concerning the application of prescribed burning to mule deer habitat in a series of publications about the habitat of the Lassen-Washoe interstate mule deer herd.[52] From 1983 to 1989, some 200,000 acres of sagebrush rangelands burned in that area. The herd ranges from Lassen County in northeastern California into Washoe County in northwestern Nevada. Loft did not differentiate between wildfires and prescribed burns in this figure, but the tone of the publications is *against* prescribed burning. The authors correctly ascertained that repeated burning at lower elevations resulted in loss of shrubs and cheatgrass dominance. They pointed out that mule deer can be forced to utilize cheatgrass, but it makes a very unsatisfactory winter forage that is lost with even a slight snowfall. The authors failed to note that grazing management, reduction in livestock numbers, and burning had restored the highland portions of the Lassen-Washoe interstate area to virtually perennial grasslands. The Shinn Peak uplands, a vast highland area dominated by perennial grasses, burned in 1974 and again in 1994. North of Observation Peak in the Buckhorn Road area there are highlands where mountain big sagebrush–antelope bitterbrush/bluebunch wheatgrass communities are being invaded by western juniper. The antelope bitterbrush plants are old and their reproduction is very limited. The perennial grass understories are excellent. Some wildlife managers think that domestic livestock are damaging the antelope bitterbrush plants and producing their generally unhealthy appearance and lack of browse production, but it is more likely that the perennial grasses that have increased with grazing management are the culprits. This is the type of site that has been renewed through prescribed burning. The decadent bitterbrush will probably be lost if the site is burned, but if the site is not burned before the western juniper trees become dominant, all browsing and grazing resources will be lost. Loft and Menke suggested that these conversions (i.e., changes brought about by burning) benefit only livestock. They do represent an excellent grazing resource, it is true, but they also probably reflect, in terms of protection from accelerated erosion and in species diversity, the pre-European landscape. Loft and Menke also pointed out that there are worse things than cheatgrass invading repeatedly burned low elevation ranges. Medusahead (*Taeniatherum caput-medusae* subsp. *asperum*) is poised to dominate millions of acres in the western Great Basin that are currently occupied by cheatgrass. In the Lassen-Washoe interstate mule deer range, medusahead will share dominance with yellow star thistle (*Centaurea solstitialis*) and Mediterranean sage (*Salvia aethiopis*). If the antelope bitterbrush at lower elevation sites is to be restored, the effort will have to involve artificial weed control and seeding technology.

In the 1970s, when prescribed burning enjoyed an unprecedented revival as a management tool, James Blaisdell's research indicating that the *intensity* of fires governed the level of antelope bitterbrush sprouting was very popular. Design the correct prescription, and decadent antelope bitterbrush stands could be renewed and released from competition. Back in the 1960s, however, Driscoll found no relation between wildfire intensity and antelope bitterbrush sprouting. He did find great variability in sprouting in different areas of wildfire burns, which he attributed to surface soil and aspect differences.[53] He pointed out, for instance, that in certain topoedaphic situations, plants whose main stem was burned all the way to the soil surface sprouted, while in others, plants with only the leaves and fine stems burned did not sprout.

Carlton Britton and Forrest Sneva tested the effect of clipping and burning on antelope bitterbrush in eastern Oregon using erect plants in a western juniper–big sagebrush community and low-growing plants in a ponderosa pine woodland.[54] More sprouting occurred from cutting than from burning, and more of the low-spreading than the erect plants sprouted. Blaisdell and Mueggler had suggested that the unusually high sprouting they observed from an August burn may have been caused by the large amount of water they used to control the experiment. This was actually tested in the Oregon experiments and found to have no effect. The final conclusion of these studies was that antelope bitterbrush does not sprout abundantly after fires. However, the authors also pointed out something we mentioned above: seed germination is the key to antelope bitterbrush regeneration. Especially in ponderosa pine woodlands, fires create ideal conditions for rodents to cache antelope bitterbrush seeds (see Chapter 7).[55]

Many scientists and resource managers continue to believe that the correct prescription will overcome the detrimental aspects of burning antelope bitterbrush stands in controlled burns. If you look at the published results of such burns, however, it should be readily apparent that there is great site and ecotypic variability in the response of *Purshia* species to fire. This does not mean that the various prescriptions are inherently wrong, but it does mean that prescribed burning of antelope bitterbrush is an extremely imprecise tool.

In the 1950s, the fact that some bitterbrush communities responded positively to fire was recognized by resource managers. Bitterbrush communities that were becoming or had become old, decadent, and unproductive were looked at as habitats that could be reinvigorated by the use of prescribed fire. The problem facing resource managers then and now is the many variables that play a role in fire-bitterbrush relations. As traditional wildlife management has preached, no two environments are alike, and each environment must be managed according to its own characteristics. Habitat managers must learn to distinguish between fire characteristics that are destructive to bitterbrush ecology and those that complement bitterbrush ecology.

Charles Driver attempted to synthesize the knowledge of wildfire–antelope bitterbrush relations into a basic principle. "How," he asked, "can a dominant plant species of a pristine climax vegetational community, with a known history of frequent fire occurrence, not express characteristics that insure its survival and perpetuation?"[56] The previous paragraphs' comments about the site specificity of natural resource management should make it clear, however, that universal statements concerning fire's role in *Purshia* communities are impossible.

In a paper published in 1980 Driver et al. established a very significant point in their conclusions concerning bitterbrush when they claimed that "*antelope bitterbrush has the ecological amplitude to function as a pioneering species following drastic disturbances within the ecosystem*" (emphasis added). The underlying problem with Driver's quest to establish antelope bitterbrush as a plant adapted to and regulated by fire is not that fire played a major, repetitive role as a stand renewal agent in precontact ponderosa pine woodlands. The pertinent question is how dominant a role antelope bitterbrush played in these woodlands. Antelope bitterbrush dominance as expressed early in the twentieth century may have been a product of postcontact changes induced by grazing, fire suppression, initial reductions in native browsers, and, potentially, climatic change. Current depletions in antelope bitterbrush dominance may also reflect the above plus the impact of alien plant colonization. Antelope bitterbrush may never have been a monotypic, landscape-dominant species under precontact conditions.

An unfortunate aspect of this lengthy discussion of *Purshia*-fire relations is that it applies almost entirely to antelope bitterbrush alone because desert bitterbrush and cliffrose are largely ignored in the literature. If the sprouting characteristics of antelope bitterbrush are inherent, they undoubtedly came from the plant's phylogenetic relations with the other two sprouting *Purshia* species.

Chapter Eleven

The Role of Nitrogen

During the 1970s we were experimenting with herbicidal control of western juniper (*Juniperus occidentalis*) on the Modoc Plateau in northeastern California, the area where the Devil's Garden mule deer herd had a spectacular population crash during the 1950s and 1960s. The herd had been depending on extensive stands of antelope bitterbrush that had reached senescence with limited or no seedling recruitment. Dillard Gates and Thomas R. Bunch of Oregon State University visited the western juniper control plots and, during the course of their review, asked what would replace the western juniper trees. We suggested that an antelope bitterbrush / bunchgrass community would be most desirable for the site. Gates expressed doubt that antelope bitterbrush would grow there, and this led to an attempt to define the types of soils to which antelope bitterbrush is adapted.

Everyone who was present contributed observations on sites where antelope bitterbrush occurs naturally. The results were near chaos. Antelope bitterbrush, we discovered, grows on deep alluvial soils and scabby, eroding slopes with virtually no soil. There are landscapes characterized by antelope bitterbrush, such as the ponderosa pine/antelope bitterbrush woodlands of central Oregon, and distinct small islands of antelope bitterbrush, apparently made safe from wildfires by highly erodible soils that do not support understory vegetation. We never arrived at a definitive definition of what constitutes a typical antelope bitterbrush site on that walk through the juniper woodlands of the Devil's Garden, but the discussion fueled our curiosity concerning the soil relations of antelope bitterbrush.

R. B. Ferguson began pioneering experiments in the population ecology of antelope bitterbrush stands in southern Idaho during the 1950s; we have extensively cited his studies in other chapters. This research raised the same soil-plant relationship question that we later encountered on the Modoc Plateau. In southern Idaho, antelope bitterbrush is found on sites with deep alluvial soils and on sites with very thin topsoil and a substratum of coarse decomposed granitic rock (mostly granodiorite and quartz monzonite), and on other soils in between.[1]

Ferguson collaborated with the soil scientist J. O. Klemmedson to determine bitterbrush seedlings' requirements for nutrients and moisture. They used the classic pot-testing methodology developed by Jenny at the University of California at Berkeley.[2] Initially they pot-tested with antelope bitterbrush seedlings grown in granitic soil collected at depths of 0–6, 6–16, and 16–36 inches. As expected, shoot and root weights of seedlings grown in the surface soil exceeded those of seedlings grown in the subsurface soils. Adding nitrogen caused seedling weights to decline in the surface soil and increase in the soil from the deepest layer. Later the study was expanded to include six granitic soils, and the authors concluded that antelope bitterbrush seedlings showed either a negative response or no response to nitrogen enrichment. Yields of antelope bitterbrush seedlings declined when the nitrogen in plants reached about 2 percent.[3]

Klemmedson and Ferguson continued this line of research by using the same set of soils and examining the influence of sulfur and potassium on antelope bitterbrush growth and nitrogen content.[4] Their results indicated that sulfur rather than potassium was the deficient element for growth in soils derived from granitic rock. Adding nitrogen, phosphorus, and potassium without adding sulfur actually reduced the growth of antelope bitterbrush seedlings in some cases. Seedlings grown in soils in which nitrogen enrichment decreased dry matter yields had elevated levels of total nitrogen and high accumulations of ammonium nitrogen and nitrate plus nitrite nitrogen. Accumulation of inorganic nitrogen was highest in the roots, next highest in the stems, and least in the leaves.

Before Klemmedson and Ferguson did their mineral nutrition research experiments, R. F. Wagle, a doctoral candidate at the University of California at Berkeley, conducted extensive trials with both antelope and desert bitterbrush soils and seedlings.[5] Wagle, who was obviously influenced by Hormay and Jenny in his studies, conducted the classic pot-test replacement series experiments:[6]

1. Control, no nutrients added
2. N + P + KS added (both K and S added as potassium sulfate)
3. P + KS added
4. N + KS added
5. N + P added

Wagle used soils from an antelope bitterbrush site at Flukey Springs in northeastern California (part of the range of the Devil's Garden mule deer herd) and from a desert bitterbrush site near Bishop, California. The Flukey Springs soil was a reddish brown loam weathered from basalt; the other soil was nearly pure pumice collected at an elevation near 8,000 feet. For bioassay plants he used the classic "Romaine" lettuce (*Lactuca sativa*) and antelope and desert bitterbrush. Wildland managers usually have little respect for data produced from trials with lettuce, but the reactions of this species in terms of leaf morphology and color are well documented for various mineral deficiencies. Nitrogen was added at the rate of 200 pounds per acre as ammonium nitrate, phosphate at 300 pounds per acre monocalcium phosphate, and potassium at 200 pounds per acre as potassium sulfate.

In the basalt-derived Flukey Springs soil, phosphorus deficiency limited shoot growth but not root growth of the indicator plants. Root growth was not significantly influenced in this soil by other nutrients (P, K, S). In the pumice soil nitrogen was the limiting nutrient. When the growth of lettuce was compared with that of antelope bitterbrush on the Flukey Springs soil, it became obvious that the bitterbrush had a source of nitrogen that was not available to the lettuce. This was attributed to the presence of nodules on the roots of the bitterbrush plants. The pumice soil either did not have the inoculum to form nodules or contained something that inhibited nodule formation. This work was conducted just as soil and plant scientists were beginning to realize that symbiotic nitrogen fixation by nonleguminous shrubs is a very significant factor in the ecology of wildland communities.[7] The fact that some nonleguminous shrubs form root nodules had been known in European forestry for a long time and in North America since early in the twentieth century.[8]

The possibility that *Purshia* species could symbiotically fix nitrogen created a new ecological perspective. Almost lost in the vast array of communities where antelope bitterbrush occurs are the numerous times it is a pioneering species on raw soils. B. A. Dickson and R. L. Crocker studied a chronosequence of soils and vegetation near Mount Shasta in northern California formed by repeated mudflows on the flank of the volcanic cone.[9] The sites ranged in age from 27 to 566 years. The older sites were occupied by ponderosa pine woodlands. Among the pioneer woody species on the raw volcanic debris was antelope bitterbrush. This same pioneering capacity on raw soils is apparent on road cuts on numerous highways in the West.

Wagle and Vlamis suggested that mineral nutrition might play a role in the apparent decadence of the antelope bitterbrush population in northeastern California.[10] Subsequently, many resource managers and scientists looked at decadent antelope bitterbrush stands and wondered if commercial fertilizers could be used to enhance their productivity. There are two related but separate problems with such decadent stands: the productivity of browse is very low, and there is no seedling recruitment. Obviously, if no leaders are produced for second-year flowering and seed production, there will be no seedling recruitment. The processes of seed germination and seedling establishment are very different from those of leader growth of established plants.

B. Leman and C. C. Pittack conducted field fertilization trials in eastern Washington using 130 pounds per acre of nitrogen applied to individual shrubs as a 30-10-0-6 compound.[11] They measured a significant increase in available browse production on antelope bitterbrush plants classed as mature, but not on decadent plants. M. A. Bayoumi and A. D. Smith studied the response of native stands of antelope bitterbrush to applications of nitrogen and phosphorus fertilizers alone and in combination.[12] Nitrogen applied at 30, 60, 96, and 149 pounds per acre to individual plants resulted in statistically significant increases in twig growth, seed production, and crude protein percentages.

Tiedemann carried out a quite complex fertilization experiment on antelope bitterbrush stands in Washington State.[13] He applied annually nitrogen, phosphorus, and sulfur at rates of 76, 29, and 10 pounds per acre, respectively. He also determined biomass and cover of herbaceous vegetation. Some of the fertilized antelope bitterbrush plants were protected from browsing. There was considerable variability among years in the amount of twig growth, but there were no consistent differences among treatments. He attributed his failure to obtain significant differences to the age of the stand. Most of the plants were overmature or decadent. He did note that mule deer preferred the browse of antelope bitterbrush fertilized with nitrogen, even though he could not show consistent increases in crude protein attributable to that fertilization. Tiedemann concluded that antelope bitterbrush may be a true pioneer species with a very low demand for nitrogen. He did not mention the potential of antelope bitterbrush to fix nitrogen symbiotically.

Wagle and Vlamis had R. V. Bega, a plant pathologist with the USDA, Forest Service, look at the root nodules they found on antelope and desert bitterbrush plants. Bega tentatively identified the organism in the nodules as an actinomycete. Since then, nonleguminous angiosperms that engage in root nodule symbiosis with actinomycetes have been found (as of 1979) in 14 genera representing 290 species, of which 145 have been reported to bear root nodules.[14] In the western United States, 8 genera with 88 species have been reported to bear root nodules. The genera and number of nodulated species are as follows: *Alnus*, 6 (8 genera); *Elaeagnus*, 3 (5); *Shepherdia*, 2 (3); *Myrica*, 3 (6); *Ceanothus*, 25 (52); *Cercocarpus*, 3 (10); *Dryas*, 2 (2); and *Purshia*, 2 (2). E. D. McArthur et al. reported that some *Cowania stansburiana* have naturally nodulated roots.[15] Since these lists were published several other genera have been added, but the fact of significance to this discussion is that all of the *Purshia* species have been reported to bear root nodules.

Klemmedson suggested that the ecological role of nonleguminous shrubs in securing unstable surfaces and promoting succession to more productive cover probably equals the utilitarian values more commonly mentioned by land managers.[16] This may be true for ponderosa pine/antelope bitterbrush woodlands, but does it apply to the critical winter habitat for mule deer, the lower elevation antelope bitterbrush–big sagebrush communities that have the most utilitarian value? R. B. Farnsworth et al. provided a very comprehensive review of the implications of symbiotic nitrogen fixation by desert plants.[17] They listed three important factors controlling the ecological significance of nitrogen fixation in desert environments: (1) the magnitude of nitrogen fixation, (2) the extent of nodulation in nonleguminous plants, and (3) the factors affecting nodulation.

Rose family genera such as *Purshia* and *Cercocarpus* are the most widely distributed actinomycete-nodulated shrubs in sagebrush communities of temperate desert environments. Actinomycete-nodulated shrubs appear to be rare or absent in shadscale (*Atriplex confertifolia*) communities in salt desert environments.[18] The warm deserts of the Southwest are nearly devoid of actinomycete-nodulated plants. Desert bitterbrush is found only at elevations above 3,000 feet in the Mohave Desert.[19] The lower elevation communities of the Mohave Desert contain numerous species of leguminous shrubs.[20]

Symbiotic nitrogen fixation is frequently cited as among the physiological attributes that permit actinomycete-nodulated species to colonize bare or disturbed sites in pioneer or seral roles of succession. Among the very first ecological papers that discussed the importance of nonleguminous shrubs as nitrogen fixers were some describing studies of vegetation development on raw glacial soils following glacial retreat at Glacier Bay, Alaska.[21] The shrub in question in these studies was *Dryas drummondii,* which belongs to the same tribe of the rose family as *Purshia.* One of the authors of this research was G. Bond of the Botany Department at the University of Glasgow, a Fellow of the Royal Society who later made important contributions to documenting symbiotic nitrogen fixation by nodulated antelope bitterbrush plants. The capacity for nitrogen fixation gives these plants a competitive advantage and constitutes a positive factor in ecosystem development.[22]

After Wagle and Vlamis first noted that the roots of certain antelope bitterbrush plants were nodulated, several papers confirming and expanding on this relationship were published. S. R. Webster, C. T. Youngberg, and A. G. Wollum, members of the Soils Department at Oregon State University, were the first to confirm that the nodules actually fix nitrogen.[23] Their research was reported in the international journal Nature. Youngberg later became a noted expert on forest and rangeland soils. The team grew antelope bitterbrush seedlings in a sandy soil collected from a ponderosa pine/antelope bitterbrush woodland in central Oregon. After the seedlings became nodulated, the nodules were excised and exposed to isotopic nitrogen (^{15}N) and then analyzed by mass spectrometry.

J. H. Becking, of the Institute for Atomic Sciences in Agriculture at Wageningen in the Netherlands, published a series of papers on nonleguminous shrubs with root nodules.[24] In classical bacteriological experiments he sterilized the surface of nodules from *Alnus glutinosa,* determined that the suspension made from the nodules was sterile except for the endophyte organism, and inoculated sterile seedlings which then nodulated.[25] He was not able to find an agar medium that supported growth of the purified organism. He is credited with being the first to identify the organism that forms the nodules and classifying it as *Frankia purshiae*.[26] Another species of *Frankia* was identified as forming nodules on the roots of mountain mahogany (*Cercocarpus*) plants.[27] R. J. Krebill and J. M. Muir used histological analysis to characterize the root nodules of antelope bitterbrush from plants collected in Idaho and Utah. They identified *Frankia purshiae* Becking from hypertrophied parenchyma cells in the cortex of the nodules.[28]

Antelope bitterbrush became the subject of research in distant Scotland when Professor Bond became involved. He was interested in shrubs of the Dryadeae (Engler) tribe of the rose family, especially the genus *Dryas*.[29] Eamor Nord and C. T. Youngberg supplied Bond with seeds of antelope bitterbrush, and he conducted the experiments in Scotland. He found that nodulated plants of antelope bitterbrush grew satisfactorily in nitrogen-free water cultures for 16 months after sowing. The fixation per unit dry weight of the antelope bitterbrush nodules was quite low compared with nonleguminous nodules, but the dry weight of the nodules constituted an astonishing 11 percent of the weight of the 16-month-old plants. Bond admitted that the growth of the seedlings was probably unfavorably influenced by very poor light conditions.

David Dalton and Donald Zobel conducted detailed field studies of nitrogen fixation at five sites in the pumice soil region of central Oregon in ponderosa pine/antelope bitterbrush communities.[30] They used the acetylene reduction method to assay nodule activity in both field and greenhouse plants. The maximum rates of fixation were observed at 74°F in greenhouse tests, but field measurement showed that soil temperatures seldom exceeded 64°F. In the greenhouse, the rate of fixation was greatly reduced at this lower temperature. Fixation of excised nodules was linear for 5 hours and then slowly declined, finally ceasing after 19 hours. Nodule activity declined in water-stressed plants and ceased in plants with xylem pressure potentials below -25 bars.

Nodule activity at the field sites began in mid-May or early June when soil temperatures at 8 inches increased above 53°F. Activity started later and remained lower under lodgepole pine (*Pinus contorta*) canopies because of the colder nature of these sites. Nodule activity at all sites was greatest in June and July, then declined in late July with the onset of moisture stress. Nodule activity always declined at night, and the decline became more pronounced as moisture stress increased. Dalton and Zobel found that only 46 percent of the antelope bitterbrush plants in the field were nodulated and suspected that the distribution of the endophyte in the soil might limit nodulation. When they examined antelope bitterbrush seedlings from old seed caches, they often found nodules on only one or two seedlings out of several in the cache. They attributed this low rate of occurrence to unfavorable soil moisture and temperature conditions. They estimated the rate of nitrogen accretion at the five pumice sites to be only 0.057 kilograms of nitrogen per hectare per year. The yearly input of nitrogen from rainwater was estimated at 0.25–0.77 kilograms per hectare per year.

Dalton and Zobel concluded that symbiotic nitrogen fixation by nodulated antelope bitterbrush plants did not appear to be an important factor in nitrogen budgets for plant communities inhabiting the pumice soils of central Oregon, which were in any case very low in nitrogen. They noted that Wagle and Vlamis had indicated that nodulation was common on soils derived from basalt in California but absent from volcanic tephra soils. Essentially, the biological significance of extremely low amounts of nitrogen in semiarid wildland communities is not well understood.

Youngberg and Wollum discussed the variation in nodulation among various types of nonleguminous shrubs.[31] Red alders (*Alnus rubra*) are almost always nodulated on sites varying from abandoned railroad grades to rich, fertile soils. Snowbush (*Ceanothus velutinus*) is similar to antelope bitterbrush in that it readily nodulates in some soils and not in others. Wollum et al. showed that the rate of nodulation of snowbush following fire or logging of Douglas fir (*Pseudotsuga menzesii*) forest depended on the age of the timber at the time of stand renewal.[32] Snowbush is not tolerant of shade and largely disappears from older Douglas fir stands. Snowbush will establish on older stand sites after they are logged or burned, but the plants will largely be void of nodulation. The lack of nodulation is apparently due to lack of endophyte inoculum. Could the same be true for antelope bitterbrush stands?

In the case of some nonleguminous shrubs the endophyte inoculum appears to be highly specific, although inoculate *Shepherdia* has been crossed with endophyte inoculum obtained from *Hippophae*, a different genus within the same family.[33] On the other hand, there are reports of incompatibility between the endophytes from species of the same genus (*Myrica*) and ineffectiveness at fixing nitrogen within another genus (*Alnus*).[34] All this led Youngberg and Wollum to conclude that strain, variation, and host specificity exist in nonleguminous as well as leguminous symbioses.

Timothy Righetti completed his doctorate at the University of California at Davis in 1980 studying soil factors limiting nodulation and nitrogen fixation in *Purshia*.[35] He published the part of this research concerning inoculation response in an article written with Donald N. Munns.[36] Righetti and Munns grew antelope and desert bitterbrush seedlings in 10 different soils collected from native bitterbrush sites in California (Table 11.1). Sites 2 (Cactus Flat) and 9 (Valyermo) were located in the mountains of far southern California; the others were on the eastern slope of the Sierra Nevada from Sierra Valley in the north (no. 3, Chilcoot) to Owens Valley in the south (no. 4, Independence).[37]

TABLE 11.1.
Soil Collection Sites in California

Location	Elevation (feet)	Estimated Precipitation (inches)
1. Burcham Flat	7,118	14
2. Cactus Flat	5,785	14
3. Chilcoot	5,135	16
4. Independence	5,135	12
5. Sherwin Hill	6,890	14
6. Shingle Mill Flat	5,948	26
7. Silver Lake	7,215	20
8. Truckee	5,785	31
9. Valyermo	3,738	11
10. Walker	5,298	15

From Timothy L. Righetti and Donald N. Munns, "Nodulation and Nitrogen Fixation in *Purshia:* Inoculation Responses and Species Comparison," *Plant Soils* 65 (1982): 383–396.

Sites 1 and 6 were nearly pure stands of antelope bitterbrush. Sites 2, 4, and 9 were dominated by desert bitterbrush. Sites 3, 5, 7, and 10 were mixed stands of antelope bitterbrush and big sagebrush. Site 8 was a ponderosa pine woodland with antelope bitterbrush and sagebrush in the understory. The antelope bitterbrush plants at site 5 had many characteristics of desert bitterbrush. The nitrogen characteristics of the soils of each location varied considerably (Table 11.2). Treatments consisted of a control, inoculation with crushed bitterbrush root nodules, and enrichment with nitrogen. The antelope bitterbrush seeds were collected at site 8 (Truckee), and the desert bitterbrush seeds were collected at site 9 (Valyermo). The nodules were collected from sites 1 (Burcham Flat) and 8 (Truckee), both of which supported antelope bitterbrush communities. Apparently, no desert bitterbrush nodules were used as inocula in this study.

At the end of two months of growth in the control soils, seedlings of both species were severely chlorotic in soils from site 1, which had the lowest total nitrogen percentage, and mildly chlorotic in soils from sites 4, 6, and 10. In soils in which the plants were naturally nodulated, recovery from chlorotic conditions appeared at three months of seedling age. In soil from site 4, only the inoculated plants showed recovery from leaf chlorosis. In terms of seedling dry weight, those grown on site 7 soil were much larger. This soil had the highest

TABLE 11.2.
Nitrogen Characteristics of Soils Used in Purshia *Nodulation and Nitrogen Fixation Experiments*

Location[1]	Total N (%)	NH_4-N (ppm)	NO_3-N initial (ppm)	NO_3-N in 6 weeks leaching (ppm)
1	0.008	2.84	0.53	0.98
2	0.016	8.70	0.93	6.85
3	0.037	6.99	8.30	8.85
4	0.016	5.84	1.32	10.38
5	0.031	12.23	2.73	17.66
6	0.005	3.09	0.04	3.31
7	0.250	78.50	39.32	61.59
8	0.137	23.26	8.17	26.63
9	0.056	14.50	1.22	29.80
10	0.022	6.54	0.96	5.10

[1] See Table 11.1 for location names.

Adapted from Righetti and Munns, "Nodulation and Nitrogen Fixation in *Purshia*," 1982.

levels of nitrogen and did not respond to additional nitrogen enrichment. This was the only soil in which elevated nitrogen levels appeared to inhibit nodulation. *Purshia* plants on site 2 soil were stunted and did not respond to nitrogen enrichment. The authors interpreted this to mean that some other required nutrient was missing. The *Purshia* seedlings failed to nodulate in the control treatments on five of the soils and exhibited only sparse nodulation on three others (Table 11.3).

When inoculated with crushed nodules from antelope bitterbrush roots, two of the nonnodulating soils and all of the sparsely nodulating soils produced seedlings with abundant nodulation. Inoculation increased nodule mass, total nitrogen, nitrogen content, and shoot biomass in bitterbrush plants from some, but not all, soils. In all but one soil, nitrogen enrichment completely eliminated nodulation. Of the three soils that failed to produce nodulated plants, one produced plants that responded to nitrogen enrichment, one produced plants with symptoms of severe nutrient deficiency, and the third produced plants that were not nitrogen deficient. Righetti and Munns concluded that the two species of *Purshia* are generally similar in nodulation, nitrogen fixation, and growth rate, although they may differ in their adaptability to certain soil conditions. In in-

TABLE 11.3.
Ethylene Production (mmols/Plant) by Bitterbrush Seedlings Grown in Five Soils

Site/ Inoculation	Species	
	Antelope Bitterbrush	Desert Bitterbrush
1. Burcham Flat		
Uninoculated	260	160
Inoculated	250	235
4. Independence		
Uninoculated	0	0
Inoculated	250**[1]	260**
6. Shingle Mill Flat		
Uninoculated	310	220
Inoculated	350*	200
8. Truckee		
Uninoculated	250	0
Inoculated	630* **	250**
10. Walker		
Uninoculated	220	240
Inoculated	980* **	240

[1]Symbols: * indicates a significant difference at the 0.05 level of probability between species grown in the same soil; ** indicates significant differences between uninoculated and inoculated treatments in the same soil.

Adapted from Righetti and Munns, "Nodulation and Nitrogen Fixation in *Purshia*," 1982.

oculation treatments that produced the sparsest nodulation, additional species differences appeared. Desert bitterbrush seedlings had fewer and smaller nodules. The authors interpreted this to mean that antelope bitterbrush nodules have more potential for nitrogen fixation than desert bitterbrush nodules. Remember, however, that only antelope bitterbrush inoculum was used in these experiments.

The lack of appreciable nodulation in soils from sites 2 and 9 led Righetti and Munns to try another experiment in which they sterilized the soils before growing inoculated *Purshia* seedlings to determine if a biological factor in the soil was inhibiting nodulation. Nodule number and nodule fresh mass showed

TABLE 11.4.
Nitrogen Fixation Parameters (per Plant) for Bitterbrush Seedlings Grown in Soils with Different Fertility Treatments

Species and Nitrogen Fixation Parameters	Unamended Soil	Phosphorus- and Sulfur-Amended Soil[1]
Antelope bitterbrush		
Nodule mass (mg)	183.0*[2]	365.0**
Nodule number	102.0*	129.0*
Total nitrogen (mg)	13.8*	18.3
Percentage nitrogen	2.6*	2.5*
Nodule percentage of roots	4.9	10.0**
Desert bitterbrush		
Nodule mass (mg)	65.0	278.0**
Nodule number	23.0	73.0**
Total nitrogen (mg)	6.7	14.4**
Percentage nitrogen	1.0	1.9**
Nodule percentage of roots	2.3	9.0**

[1] Soil amendments: 50 ppm phosphorus as $Ca(H_2PO_4)_2$ and 80 ppm sulfur as $CaSO_4 \bullet 2H_2O$.

[2] Symbols: * indicates the parameter difference at the 0.05 level of probability in the amended treatment; ** indicates the value among species is significantly different at same level of probability.

Adapted from T. L. Righetti, "Soil Factors Limiting Nodulation and Nitrogen Fixation in Bitterbrush (*Purshia*)" (Ph.D. diss., Univ. Calif., Davis, 1980).

no improvement following sterilization of these two soils. The other soils were also sterilized before inoculation, and the results were inconsistent.

Even with a suitable endophyte population and soil moisture, optimal nodulation and nitrogen fixation may be limited by soil fertility.[38] Fertilizing soils with phosphorus and sulfur increased nodule mass, nodule number, total nitrogen in shoots, percentage nitrogen in shoots, and percentage of the seedlings consisting as root nodules (Table 11.4). Only when he enriched the growth substrate with sulfur and phosphorus did Righetti's plants approach the percentage weight of root nodules that Bond reported for antelope bitterbrush.

Larger, healthier plants could probably be produced through inoculation with compatible and superior strains of endophyte, thus enhancing nitrogen fixation.[39] Not all *Frankia* strains are equally capable of fixing nitrogen, and the endophyte at a given location may not be the most effective. Perhaps ineffective strains could be isolated and removed from natural populations.[40] The ability of a superior strain to compete with ineffective endogenous strains is unknown. If container-grown nodulated plants were transplanted to the field, one would be assured at least for the short term that desirable and superior symbiosis existed.

We believe that the widespread failures with bare root antelope bitterbrush transplants may be connected with lack of endophyte inoculation. This would be especially true of plants produced by nurseries that use sterile growing media to reduce damping-off. Several studies have indicated that establishment of symbiosis is rather slow. This known delay prompted Righetti et al. to propose that nitrogen fertilization of transplanted seedlings might be appropriate on certain soils.[41] Nitrogen fertilization also enhances the growth of cheatgrass, however, and cheatgrass competition is usually fatal for antelope bitterbrush seedlings.

Righetti and Munns also experimented with cliffrose seedlings in greenhouse tests and demonstrated nodulation and acetylene reduction.[42] Nitrogen fixation was verified with isotopic nitrogen. The seedlings were grown in a soil known to nodulate both antelope and desert bitterbrush, to which was added a suspension of crushed antelope bitterbrush nodules. Bitterbrush seedlings grew faster and reduced more acetylene than cliffrose seedlings grown in the same pot, but the specific activity of the nodule mass–to–root mass ratios was approximately the same. The authors noted that sampling only the surface soil may produce misleading results regarding the inoculation potential of cliffrose seedlings (Table 11.5).[43]

David Nelson and Patti Schuttler described the morphology and histological characteristics of cliffrose nodules.[44] Like the nodules of other members of the rose family, cliffrose nodules appear to be perennial with an indeterminate, generally dichotomous branching that eventually develops into coralloid nodule clusters up to a couple of inches in diameter. As is the case for other actinorhizal families, the nodule anatomy resembles that of the primary root, except there are few root hairs, no cap, and a superficial peridium rather than an epidermis. The endophyte invades immediately proximal to the meristem, first stimulating enlargement of the host nucleus, then causing cell enlargement.

Arizona cliffrose (*Cowania subintegra*) has also been reported to have nodulated roots.[45] Apache plume (*Fallugia paradoxa*) plants, on the other hand, have not been found with natural root nodulation.[46] This is interesting because natural putative hybrids between cliffrose and Apache plume have been reported from the field.

Plant pathologist David Nelson made the interesting observation that at the time he was writing (1982), no one had succeeded in culturing *Frankia* in the absence of its host. The failure to find a natural saprophytic phase suggests that the *Frankia* are obligate parasites. In general, obligate parasites infect and grow

TABLE 11.5.
Nodule Number (per Plant) for Six-Month-Old Cliffrose Seedlings Grown in 10 Arizona Soils with and without Nitrogen Enrichment

	Unamended		Nitrogen Amended (56 ppm)	
Soil Location	Surface (0–8 inches)	Subsurface (8–16 inches)	Surface (0–8 inches)	Subsurface (8–16 inches)
Big Springs	5	15	0	16
Desert View	0	1	0	0
Flagstaff	41	266*[1]	0**	7**
Fredonia	3	49*	0	21
Grand Canyon	2	29	0	1
Jacob Lake	4	81*	0	4**
Payson	3	6	3	7
Prescott	6	18	3	4
San Francisco Peak	29	46	8	85*
Williams	46	23	35	10

[1] Symbols: * indicates that values are significantly different at 0.05 level of probability in the surface and subsoil treatment; ** indicates the values are significantly different at the same level of probability between control and nitrogen enriched treatments

Adapted from Timothy L. Righetti, Carolyn H. Chard, and D. N. Munns, "Opportunities and Approaches for Enhancing Nitrogen Fixation in *Purshia, Cowania,* and *Fallugia,*" pp. 214–223 in Tiedemann and Johnson 1983.

better in a healthy, vigorously growing host. If nitrogen deficiency in the host enhances infection, that would be inconsistent with other cases, but perhaps it is unique to *Frankia* obligate parasitism.[47]

The biology of actinomycete nodulation of the roots of rose family shrubs was analyzed with the tools of modern molecular biochemistry through the efforts of Wolfgang Hornerlage and his associates to develop RNA probes to characterize *Frankia* strains.[48] Comparative sequence analysis of PCR-amplified and -cloned inserts from 35 *Frankia* strains confirmed the separation of these strains into specific host infection groups.

One thing is clear: symbiotic nitrogen fixation by *Purshia* species does exist. It might not exist in all soils where bitterbrush grows, and we are not yet sure of the *biological significance* of the amount of nitrogen that is fixed by *Purshia* species. There is virtually no information on the importance of symbiotic nitrogen fixation during the phenological development of bitterbrush species. Seedling establishment is generally considered the most critical stage of shrub establishment, and the experimental evidence concerning nitrogen fixation by nodulated bitterbrush plants comes from work with seedlings. We need to know a great deal more. What is the nodulation status of decadent antelope bitterbrush stands? Do the nodules die as the general physiological status of the plants declines, and if they do, does the decay and mineralization of the nodules enrich the subcanopy area with nitrogen? This could be very important in seedling establishment, especially if a nitrogen-loving invasive weed like cheatgrass could take advantage of the released nitrogen to enhance its own competition with antelope bitterbrush seedlings.

Our discussion of the mode of infection of the symbiotic organism has assumed that the inoculum is present in some soils and not in others. What if a rodent vector were involved? We have noted that virtually the entire antelope bitterbrush seed crop is collected and cached in scatter-hoards by granivorous rodents. Given the fact that granivorous rodents dig in the surface soil, the possibilities for the spread of inoculum seem obvious. Bill Longland and Jim Trent of the USDA, ARS, in Reno made some preliminary investigations of the microorganisms found in rodent cheek pouches. They were not specifically looking at antelope bitterbrush seeds and endophyte inoculum, but they did determine that the external cheek pouches of rodents have distinctive microorganism assemblages. There may be some logical reason why rodents cannot serve as vectors of the necessary inoculum, but the possibility should be investigated.

In the course of research on nodulation and nitrogen fixation, many other microorganisms have been found living in or on the roots of *Purshia.* Stephen Williams, for example, found abundant evidence of vesicular-arbuscular mycorrhizal symbionts on the roots of mountain mahogany (*Cercocarpus montanus*) and antelope bitterbrush.[49] Such infections are thought to be beneficial to the host plant because they enhance the uptake of phosphates and water. The influence of vesicular-arbuscular mycorrhizal infection on ectomycorrhizae infection and nitrogen fixation is unknown.

Many years of research on the competition between cheatgrass and the seedlings of temperate desert plants have shown that competition in this environment is overwhelmingly for moisture. However, the availability of nitrogen, often in tiny amounts by agronomic standards, is the catalyst that governs this competition. Cheatgrass thrives on nitrogen. With adequate nitrogen it can completely close seedbeds to the establishment of perennial seedlings. We developed an experimental procedure to test this nitrogen relationship by enriching soil with various forms of nitrogen (calcium nitrate, urea, or ammonium sulfate) or immobilizing available nitrogen through supplying a readily usable source of carbon for soil microorganisms (sucrose).[50] We inhibited nitrification

TABLE 11.6.
Antelope Bitterbrush Seedling Density and Height in August 1994 after One Growing Season[1]

Treatment	Seedling Emergence (%)	Seedling Height (inches)
Control	10bc	1.2de
Urea	6bc	2.0de
Calcium nitrate	15ab	3.2cd
Ammonium sulfate	0c	0.0e
Carbon	11b	7.2a
Nitrapyrin + carbon	12b	4.8bc
Nitrapyrin	13b	8.0a
Nitrapyrin + ammonium sulfate	25a	1.6de

[1]Means followed by the same letter, within columns, are not significantly different at the 0.05 level of probability as determined by Duncan's Multiple Range Test.

Adapted from J. A. Young, Charlie D. Clements, and Robert R. Blank, "Influence of Nitrogen on Antelope Bitterbrush Seedling Establishment," *J. Range Manage.* 50 (1997): 536–540.

through applications of nitrapyrin. Combination treatments included carbon plus nitrapyrin and the application of ammonium sulfate with nitrapyrin to hold the nitrogen in the form of ammonia. In the fall of 1993 we applied these treatments, plus a control, to experimental plots and placed antelope bitterbrush seed caches in each treatment. The experiments were conducted at Doyle, California, in the shadow of E. A. Nord's exclosure where so much of the early research on antelope bitterbrush ecology was done. The soils at the site are complex because they have developed on delta material where Long Valley Creek emptied into pluvial Lake Lahontan, but the primary surface soil is largely derived from decomposed granite from the adjacent eastern slope of the Sierra Nevada. When we managed to establish antelope bitterbrush seedlings in the treatments, we found striking differences in the height of the seedlings and the comparative growth of cheatgrass and other annual weeds (Table 11.6).[51]

Only 6 inches of precipitation fell at the experiment site in the winter and spring of 1993–1994. Our nitrogen manipulations did influence antelope bitterbrush seedling establishment, but the interactions are difficult to interpret. One thing that is clear is that in the plot enriched with ammonium sulfate, the antelope bitterbrush seedlings were completely swallowed by a dense sward of cheatgrass. In these arid rangeland environments, ammonium sulfate is often the most biologically effective means of nitrogen fertilization, probably because it is comparatively soluble. The two treatments with the highest percentage of emergence were calcium nitrate and nitrapyrin plus ammonium sulfate. Plots in which nitrogen was immobilized with carbon or nitrification was inhibited with nitrapyrin produced 7–8-inch-tall seedlings. The control seedlings were slightly over an inch tall, and seedlings in soil that received the best nitrogen enrichment treatment (calcium nitrate) were only half the height of those that received the carbon and nitrapyrin treatments. These two treatments and the plot treated with a combination of carbon and nitrapyrin were nearly completely free of annual weeds, including cheatgrass. Unfortunately, we did not evaluate seedling nodulation in these experiments.

In the semiarid environments where the *Purshia* species are found, plants compete for moisture above all else. Nitrogen, the essential element for plant growth that is often limiting in these environments, is the catalyst that governs this competition. This makes the nitrogen fixed by symbiotic nodules on the roots of *Purshia* species critical even if the amount of nitrogen gained by this means is small.

Chapter Twelve

Purshia Management

The management of *Purshia* species today is strongly focused on antelope bitterbrush, primarily because of its importance to mule deer. For much of the nineteenth century, however, range managers managed *Purshia* as forage for domestic livestock.

Early range research focused on livestock grazing because of the economic importance of the range livestock industry. In 1932, for example, G. D. Pickford reported that "the production of livestock is a very important phase of agriculture in [Utah], . . . hence the economic welfare of the state depends to great extent upon keeping the grazing lands fully productive."[1] As the West was settled, productive rangelands were turned into farms, and the ranges that remained were overgrazed by the growing herds of livestock (Fig. 12.1). In 1883, for example, there were an estimated 100,000 cattle and 450,000 sheep in the Utah territory.[2] Less than 10 years later, in 1891, the number of sheep had grown to 3,537,000.[3] By 1931 there were 344,000 cattle and 2,926,000 sheep in the state.[4]

In the late 1930s and early 1940s, the realization that forage was being depleted throughout the western rangelands led to experiments designed to determine the proper use of key browse species.[5] In many areas, however, stockmen and natural resource managers were not sure which plants were the key species for domestic and native grazers.[6] Arthur W. Sampson, the man who reported that antelope bitterbrush was a strong feed that produced a solid fat, was the first to consider antelope bitterbrush an important browse species.[7] Another early researcher, Joseph Dixon, studied the food habits of California deer in different regions of the state (Modoc, Lassen, Plumas, Mariposa, Fresno, and San Diego Counties) in the 1920s and early 1930s by actually observing the food selection of grazing deer. Although Dixon rated the relative importance of antelope bitterbrush to deer as "great" (very important) in Lassen County during the winter months, he did not mention the plant in any of his observations for the other areas.[8] The *Range Plant Handbook*, published in 1937, reported that bitterbrush was an important shrub on winter ranges for deer, elk, and antelope.[9]

In what may well have been a response to the debate raging between hunters and ranchers over appropriate range use, G. D. Pickford reported in 1932 that

Fig. 12.1. Heavily grazed rangeland, ca. 1900. From D. Griffiths, *Forage Conditions on the Northern Border of the Great Basin* (Bull. 15, Bur. Plant Industry, USDA, Washington, D.C., 1902). This is one of many references to the depletion of range resources by excessive grazing late in the nineteenth century. The plant communities that were destroyed by excessive grazing left behind soil resources that were exploited by the shrub-dominated communities that replaced them. The same soil resources may not be present when we try to artificially replace such shrub stands.

antelope bitterbrush density increased on grazed habitats in Utah.[10] The conflict continued nevertheless into the late 1930s and the 1940s. Among the first to address the range use debate in print were L. A. Stoddart and D. I. Rasmussen, who in 1945 published a paper titled "Deer Management and Range Livestock Production,"[11] which supported the view that livestock and deer could share rangelands. The authors did recognize the palatability of antelope bitterbrush and cliffrose to both livestock and deer, thus acknowledging that serious competition for these species could occur. Stoddart and Rasmussen also suggested that half of Utah's range had no deer population, and thus no deer-livestock competition. They used a "competition index" to prove that cattle would use only about 75 percent of the area used by deer, and would eat only 25 percent of the plants eaten by deer. The problem with this calculation, however, is that very little was known at the time about the diets of different deer herds.

In the years since, management of *Purshia* range resources has generally fallen into four categories:

1. Prevention of use by domestic livestock, thereby preserving the range for mule deer
2. Reduction of mule deer use by management practices such as decreasing mule deer numbers through hunting quotas or using fenced exclosures
3. Enhancement of stands through seedling recruitment
4. Manipulation of existing stands to enhance productivity

Reducing competition with domestic livestock has generally involved reducing the number of livestock permitted to graze on publicly owned rangelands, which may involve extensive lobbying efforts, developing and implementing grazing management systems on publicly owned rangelands, purchasing privately owned rangeland for exclusive use by mule deer, and paying incentive rewards for the management of privately owned rangelands to enhance mule deer habitat.

In the early twentieth century grazing management began to be applied routinely across the western ranges. Odell Julander and Leslie Robinette reported that in the portion of Utah where they studied deer and cattle interactions, reductions in the number of grazing permits issued by the U.S. Forest Service had reduced cattle use by 68 percent by the 1950s.[12] This practice was repeated on many rangelands, and the number of permits issued continues to decrease.

Along with the realization that the western ranges were deteriorating came awareness of the need to protect America's mule deer herds, which had declined to perilous numbers by the early twentieth century. Hunting seasons were either closed or shortened, and bag limits were enforced. Refuges were established where mule deer were protected from hunting. These measures along with predator control programs allowed mule deer herd populations to mushroom throughout the West. The result was rangelands overgrazed by both livestock *and* mule deer. This was most noticeable on winter ranges, which were generally smaller in area and received more concentrated use by mule deer.

By the late 1930s and early 1940s wildlife managers recognized that these "hot spots" were overutilized, and those habitats that were difficult to restore to full productivity were being destroyed. Soon, mule deer herd populations began to crash through winter starvation. In the early 1940s, the U.S. Wildlife Service studied a deer herd and its effects on winter browse in the Schell Creek Division of the U.S. Forest Service in eastern Nevada (27,520 acres).[13] The resulting report concluded that the mule deer densities were too high and that the herd needed to be reduced even further to protect both the winter range and the mule deer herd from permanent damage. Harvesting bucks alone would not be sufficient, and the report suggested harvesting does as well, both in this location and elsewhere where mule deer populations were too high.[14] Thus came the proposal for "antlerless" (doe) hunts to bring the population down to a size compatible with the carrying capacity of the range.

The protection of does had been a staple of mule deer management since the earliest days of the twentieth century when mule deer populations faced extinction in many parts of the West, and most hunters had never shot one (openly, at least). The senior author remembers eavesdropping on a conversation among hunters at a service station in rural Siskiyou County, California, in the early 1950s. Ralph Smith protested that he was going to feel awful strange driving down Main Street with his deer tag around an old doe's ear. One of the old-timers sitting around the stove said, "Oh, hell, Ralph, if you kill a doe you won't drive down Main Street, you'll come down the back alleys out of force of habit."

There is an often repeated story concerning the first such hunt in the Lavabeds winter ranges in Modoc County, California, where the mule deer population had greatly exceeded the range resources. Hunters became so excited at the opportunity to shoot does that a game warden had to crawl out with a white flag to rescue a wounded hunter.

The Schell Creek Range studies provided data that allow a comparison of mule deer densities over time. The U.S. Wildlife Service counted 1,230 mule deer there in the spring of 1942, and 2,009 in the spring of 1943.[15] And while the Nevada Division of Wildlife (NDOW) counted only 600 deer in the entire state in 1951, later in the 1950s a group of sportsmen on horseback counted 3,000 mule deer in an area just north of Reno—and they counted only the deer that broke to their left.[16] Inefficient counting methods and insufficient effort may have hampered the NDOW's count and skewed the results. By the late 1970s the NDOW was using aircraft (helicopters and fixed-wing craft) to record mule deer population data. Dave Mathis, an information officer with the NDOW at that time, reported in his book *Following the Nevada Wildlife Trail* that when they first started using aircraft, wildlife biologists discovered that they saw three times more deer from the air than they did on the ground.[17] At one location, for example, the biologist saw 200 deer from a pickup truck while 400 were counted from a fixed-wing aircraft and almost 700 were counted from a helicopter. In 1998 the Schell Creek mule deer population was estimated to be 10,100.[18]

In the late 1970s the NDOW radio-collared mule deer and used the more refined data they collected to estimate that their counts represented approximately 45 percent of the total population.[19] Traditional wildlife managers estimated that they saw about 33 percent of the total deer population on their ground counts, so we assume that the count efficiencies back in the 1940s were in that neighborhood as well.

Although mule deer populations have fluctuated in the past 50 years, the use of browse plants has remained fairly constant. Back in 1943, based on a count of 2,009 mule deer and an unknown number of livestock, Aldous reported that antelope bitterbrush had 51 percent utilization and that cliffrose had 38 percent utilization in the fall months. In 1997 these utilization percentages were 51 percent on antelope bitterbrush and 50 percent on cliffrose in the month of September with a mule deer count of 2,800. Livestock numbers have declined since the 1940s, but elk have increased dramatically in recent times—from an estimated 500 in the Schell Creek Range back in the 1940s, for example, to 2,355 in 1998,[20] a 470 percent increase. The fact that livestock numbers have decreased and elk numbers have increased but the figures on utilization of antelope bitterbrush and cliffrose are very similar to the 1943 figures may mean that mule deer numbers are lower at the present time. In fact, it has been suggested that the mule deer numbers are lower today than in the 1940s, but also that antelope bitterbrush and cliffrose habitats have decreased over time as a result of habitat changes such as pinyon-juniper encroachment.

We have already discussed the difficulties inherent in artificially planting *Purshia* seedlings. From a management standpoint, successful renewal of the resource starts only when the seedlings are established. *Purshia* seedlings face potential excessive utilization by rodents, jackrabbits, mule deer, and livestock. Prevention of excessive utilization by mule deer involves reducing deer numbers through manipulation of the hunting quotas, fencing, or altering the scale of the restoration project. Fencing to exclude mule deer is very expensive. It has been extensively practiced in areas of Lassen County, California (Fig. 12.2), often by using prison inmate crews to reduce the high labor cost. A 16-man prison inmate crew cost $320 per day in the late 1990s, and the material needed to construct a big game exclosure adds to the expense. These exclosures have to be maintained for at least a year to give the seedlings a chance to establish. In Lassen County the California Department of Fish and Game constructed a 700-acre big game exclosure following a wildland fire in the Bass Hill country. The site was seeded back to antelope bitterbrush to try and restore the critical mule deer winter range. The exclosure has been removed now, and 3,000 pounds of antelope bitterbrush seed are being harvested annually.[21]

Exclosures have also been used to determine which animals are doing particular types of damage to the range. In the late 1940s and early 1950s wildlife agencies began building big game exclosures to monitor the level of use of vegetation by different classes of animals (Fig. 12.3). These exclosures helped managers demonstrate to different interest groups (i.e., livestock operators and sportsmen) that livestock, and not deer, were primarily responsible for the

Fig. 12.2. Mule deer exclosure constructed on land owned by the California Department of Fish and Game near Doyle, Lassen County, to protect antelope bitterbrush seedlings from mule deer.

depleted range grasses, while too many deer had a negative effect on browse species.[22]

The scale of each restoration project influences the seedlings' chances for establishment by influencing deer and, especially, jackrabbit predation on the young plants. Studies of black-tailed jackrabbit (*Lepus californicus*) predation on crested wheatgrass seedlings, for example, showed a pronounced edge effect on the crested wheatgrass stand in relation to escape cover.[23] The farther the seedlings were from escape cover, the less the predation by the jackrabbits. Small-scale restoration projects surrounded by a sea of degraded habitat are particularly subject to excessive predation. In such cases, fencing to exclude both mule deer and jackrabbits is essential to success. It is also possible to manipulate the escape cover outside or surrounding the restoration project to reduce predation. Use of a mechanical brush beater to knock down the aerial portion of the shrubs, for example, may have no lasting effect on sprouting shrubs, but it can reduce escape cover in the short term. Jackrabbit populations are cyclical, and at low points in the cycle jackrabbit predation may be of no consequence. Conversely, at the peak of the population cycle only adequate fencing will protect seedlings. Unfortunately, no one has derived a means of precisely predicting the cycles in jackrabbit populations because crashes depend on mid-

Fig. 12.3. A three-way big game exclosure built in the 1950s to aid in assessing utilization of important browse species by different classes of herbivores (i.e., deer and cattle).

winter weather (snow) conditions.[24] A field census of jackrabbit densities should be part of every restoration plan.

We discussed the relationships of granivorous rodents to antelope bitterbrush in Chapter 7. Here we discuss such relations as they affect management decisions. Granivorous rodents not only influence the success of restoration projects through their seed predation, they can also have a significant effect on seedling survival. Our research on seedlings in a sagebrush–antelope bitterbrush community in northeastern California found 87 percent predation by three species of rodents.[25] In the spring of 1998, after five consecutive years of above average precipitation, the same location in northeastern California had an excellent year in terms of the number of antelope bitterbrush seedlings that emerged. Most of these seedlings were in caches. After one month of observation, however, we found 32 percent cache survival and only 17 percent seedling survival. Like jackrabbits, rodent populations are cyclical, and resource managers can use live-trapping to determine rodent densities at any given time.

Excessive browsing of antelope bitterbrush seedlings and young plants by domestic livestock is a complex issue involving such variables as season of use, stocking levels, and alternative forage. Relatively pure stands of antelope bitterbrush without associated perennial grasses are an open invitation for excessive utilization by domestic livestock. If cattle are given access to relatively pure stands of bitterbrush, especially late in the summer after the herbaceous vegetation has dried, excessive browsing is almost assured unless their numbers and the duration of grazing are limited. An old cowboy in Tuscorara, Nevada, told us that whenever they came up short on their fall gathering, they always looked in bitterbrush patches for the missing cattle.

Wildlife managers have long been reluctant to plant perennial grasses along with bitterbrush, fearing that mixed plantings would lead to excessive competition at the expense of the shrubs.[26] In fact, however, as should by now be clear, such grasses are necessary for the stand's survival. The grazing of associated herbaceous vegetation in browse stands can *limit* competition for water and nutrients and, most important, change the fuels available for wildfires. It is counterproductive to invest in the restoration of antelope bitterbrush communities only to have them destroyed by wildfires before the browse can be utilized. As we discussed earlier, the critical fuel load characteristic is the abundance and productivity of cheatgrass in the understory. Antelope bitterbrush plants themselves cannot adequately suppress the productivity of cheatgrass herbage; that requires a perennial grass. In recent years we have witnessed significant increases in perennial grasses in certain habitats due to changes in grazing management, particularly season of use and decreased livestock numbers permitted. In these habitats, the increase in perennial herbaceous material has also increased the fuel load, thus increasing the chances that a wildfire will ignite and spread. There is a definite need for research on the spatial distribution of shrubs and grasses in restored browse communities. Perhaps antelope bitterbrush and perennial grass plants should be physically separated in the planting design. It is obvious that any large expanse of antelope bitterbrush needs fuel breaks to aid in wildfire suppression. These fuel breaks need to be dominated by perennial grasses, and the perennial grasses in turn need to be periodically grazed to reduce the accumulation of flammable litter.

A photograph we took in Cave Valley, Lincoln County, Nevada, in December 1999 illustrates some of the dilemmas faced by wildlife habitat managers (Fig. 12.4). A broad old alluvial fan at the base of the Schell Creek Range supports an excellent stand of antelope bitterbrush. The plants are old, quite heav-

Fig. 12.4. Nearly senescent stand of antelope bitterbrush showing no seedling recruitment in Cave Valley, Lincoln County, Nevada, with the Schell Creek Range in the background. The site has an excellent understory of native perennial grasses.

ily browsed, and often high-lined by browsing into umbrella-shaped growth forms. Smaller dead shrubs are apparent. The understory is an excellent stand of native perennial bunchgrasses with occasional cheatgrass plants. If the site burns in a wildfire or prescribed burn, it will probably be converted to a perennial grassland with minimal shrubs. If the site does not burn, the senescing antelope bitterbrush plants, which are not reproducing, will gradually disappear, especially if grazed by cattle late in the fall when their browse represents the only source of digestible protein. Complete removal of domestic livestock would increase the fire hazard by allowing a greater accumulation of herbaceous fuel and probably would do little to prolong the life of the old antelope bitterbrush plants. Heavy grazing of the perennial grasses in the spring *might* favor antelope bitterbrush seedling recruitment, *if* the old plants produce sufficient seed. Complex interacting problems in antelope bitterbrush management such as these are repeated in various combinations across the entire range of the *Purshia* species.

The escalating costs and inherent dangers associated with complete suppression of fire have brought directives from the highest levels of natural resource administration to use prescribed burning to reduce the risk of catastrophic wildfires. The influence of such prescribed burning programs on *Purshia* species is perhaps the most important issue facing mule deer habitat managers today. The Cedar Creek project south of the Warner Mountains in northeastern California provides an example of the interactions of prescribed burning programs and mule deer habitat.

The Cedar Creek basin was once excellent rangeland for both domestic livestock and mule deer. Mountain big sagebrush/bluebunch wheatgrass communities predominated, with occasional aspen and curlleaf mountain mahogany stands at higher elevations. At the beginning of the twentieth century, western juniper stands were restricted to rimrock and talus slopes in the general basalt flow topography. By the 1970s western juniper dominated virtually the entire basin, and shrubs and herbaceous species had been purged from the understory. In the late 1980s, land managers proposed prescribed burns to reduce the dominance of western juniper and also to reduce the chance of catastrophic wildfires in the maturing stands of juniper.

If the suppression of wildfires or the lack of wildfires resulting from the reduction of herbaceous fuels by grazing had been the cause of the increase in western juniper, then the reintroduction of prescribed fire should have improved the area for grazing. What actually happened, however, was that the prescribed burn also destroyed portions of the remnant stands of antelope bitterbrush and curlleaf mountain mahogany. Mule deer habitat managers were faced with an immediate decline in browse resources. Obviously, the argument can be made that these remnant stands were in the process of being crowded out by the encroaching junipers anyway, and it was only a matter of time before they would have been lost to this invasion. The prescribed burn should have paved the way for succession to proceed to a point at which new stands of antelope bitterbrush would be established. But the short-term result was the destruction of an already sparse and overutilized browse resource. This same scenario, in various forms, is going to be a consideration in management decisions concerning mule deer habitat and the use of prescribed burning throughout the West.

What are the management alternatives? Perhaps the prescribed burns could be made small to reduce the gross negative impacts on existing habitats. There is a critical practical minimum size in prescribed burns, however, especially in juniper woodlands. The fire has to be large enough to generate enough energy to be self-perpetuating. Furthermore, small openings in degraded habitats are predetermined to be excessive sites of seedling predation, which often prevents the return of woody species. Active restoration programs to establish browse species populations artificially following the prescribed burn are one way to ensure continuity in browse resources. The time required for antelope bitterbrush populations to produce amounts of browse sufficient to support browsing suggests the feasibility of using faster-growing species such as desirable ecotypes of big sagebrush. Perhaps a "nurse" crop of big sagebrush should be planted in association with antelope bitterbrush restorations as a transitional browse resource following prescribed burns.

In the 1950s a very popular range improvement treatment for big sagebrush/bunchgrass rangelands consisted of aerial applications of the herbicide 2,4-D to reduce shrub dominance and release native grass species. Wildlife managers roundly condemned this practice as having severe negative impacts on browsing resources by killing antelope bitterbrush plants.[27] Not everyone agreed that the herbicide was harmful to bitterbrush. D. N. Hyder and F. A. Sneva experimented with selective removal of big sagebrush plants from antelope bitterbrush stands with carefully timed applications of 2,4-D.[28] Apparently, there were no follow-up studies to determine if reducing big sagebrush plant densities enhanced the growth of antelope bitterbrush plants. Considering what we now know about the importance of big sagebrush in the winter diet of mule deer, we are certainly not advocating the application of 2,4-D to antelope bitterbrush stands, but long-term studies of population dynamics and productivity of stands where big sagebrush was selectively removed would be very interesting. Apparently, no one has investigated the influence of other shrubs on *Purshia* species' persistence and productivity in mixed communities.

Another aspect of browse species management that has received scant or no attention from researchers is the possibility that excessive browsing may have very negative impacts on the genetic pool of the preferred species. Browsing animals exhibit obvious preferences for certain plant ecotypes over others. Has past excessive utilization by mule deer populations selected for relatively nonpreferred forms of antelope bitterbrush?

We earlier discussed fertilization of shrub stands as a way to enhance browse production and manipulate animal preferences. Experiments have proved that production, nutrient content, and utilization can be changed through fertilization, but practical applications involving artificial fertilization have not followed. Cost-effectiveness is an obvious factor.

The success of the wildlife preserves employed to enhance mule deer populations early in the twentieth century is difficult to evaluate because they often represented only a fraction of the herd's yearly range. Lack of management outside preserves may have overshadowed any benefits gained from lack of hunting on preserves. Wildlife preserves as management tools had generally fallen into disfavor by the mid-1900s, only to enjoy a revival late in the century when game management agencies were able to obtain outside funds to buy property. Examples include the Mountain Lion Initiative in California and the activities of such organizations as The Nature Conservancy throughout the West. One frequent problem with this type of acquisition is that funds for managing the acquired resource are often not included.

When working ranches are acquired for mule deer habitat and the farming operations are stopped, the result is not always beneficial for the deer. Managers often hear that "the mule deer were better off while the ranch was still being farmed." Two aspects of this type of habitat management are appropriate to our discussion. First, if some form of *restoration* of browse resources is not practiced on such damaged habitats, any positive influence on mule deer populations will be a long time coming. In fact, the preservation will be likely to have *negative* influences on mule deer habitat through such mechanisms as increased herbaceous fuel loads in browse stands. Second, the failure of deer to thrive on working ranches converted to game preserves illustrates how dependent mule deer populations in areas with degraded browse resources have become on agricultural crops, especially alfalfa (Fig. 12.5). Few big game managers advocate hay farming to support mule deer populations, but the importance of agricultural crops in meeting the nutritional requirements of mule deer needs to be considered when habitat preserves are established.

Topping was once proposed as a management treatment to increase the browse productivity of overmature antelope bitterbrush plants. Ferguson and Basile first proposed this treatment in Idaho in the 1960s and then later tested it at several locations, including the Modoc National Forest in northeastern California.[29] Overmature antelope bitterbrush plants from 5 to 7 feet tall were cut off 3–4 feet above the ground. The purpose was to increase twig production on the limbs remaining below the cut. Before the cutting, many of the twigs produced on the limbs above the cut height were unreachable by mule deer. Topping initially stimulated twig production at all locations where it was tested, although the results were quite variable (Table 12.1). Within four years of the topping, however, production was not significantly different from the control plants. This type of labor-intensive treatment is very expensive (in the 1960s, $10–35 per acre). The growth response to cutting is apparently mediated by a hormone, but it seems that no one has ever tried to enhance twig production of antelope bitterbrush plants by topical applications of plant growth regulators. That also, however, would be an expensive and labor-intensive treatment.

Perhaps the most difficult issue in the management of *Purshia* stands involves the question of why so many antelope bitterbrush stands established 80–100 years ago (in the early 1900s) are currently experiencing inadequate seedling recruitment. At that time the western ranges were excessively overgrazed, but these overgrazed ranges also experienced unprecedented events.

Fig. 12.5. Mule deer depredation of agricultural fields can be heavy, as shown here at the Buffalo Meadows ranch in northwestern Nevada.

The hard winter of 1889–1890, for instance, reduced livestock numbers by almost 90 percent in northern Nevada,[30] and ungulate wildlife populations were also very low as a result of promiscuous hunting. It is very difficult, and perhaps impossible, to make absolute cause-and-effect statements as to why bitterbrush stands became established at the turn of the century but are now experiencing difficulties in recruitment.

Perhaps the low number of ungulate grazers immediately following the 1889–1890 winter favored the establishment of bitterbrush. Other factors working in favor of establishment in 1890–1910 include (1) reduced competi-

TABLE 12.1.
Mean Twig Growth of Topped and Untopped Bitterbrush Plants over Four Growing Seasons

Location	Year	Topped (inches)	Control (inches)
Boise, Idaho	1966	389*	75
	1967	358	155
	1968	186	111
	1969	166	145
Sawtooth, Idaho	1966	296*	48
	1967	187*	45
	1968	95*	32
	1969	188	112
Modoc, Calif.	1966	541*	225
	1967	768*	332
	1968	183	120
	1969	239	145

* Asterisk denotes production significantly different from the control.

Adapted from R. B. Ferguson, "Bitterbrush Topping: Shrub Response and Cost Factors" (Res. Paper 125, USDA, Forest Service, Ogden, Utah, 1972).

tion from perennial grasses because of extreme overgrazing, (2) reduced wildfires due to lack of herbaceous fuel, (3) promiscuous fall burning by stockmen, and (4) lack of exotic annual grasses such as cheatgrass, which had not yet been introduced. Many photos of western rangelands taken in the late 1800s show areas void of all herbaceous vegetation, with only a bush here and there. Is it possible that this lack of herbaceous material along with the severe losses of livestock and low numbers of native wild ungulates produced a vacuum into which such woody plants as *Purshia* were able to move? Rangelands that were first overgrazed and then left ungrazed could have been exploited by woody plant seedlings unrestrained by domestic or wild herbivores.[31] With the increase in woody plants also came a decrease in wildfires because the fine fuels needed to carry fires were lacking. Fire-sensitive plants such as sagebrush and antelope bitterbrush benefited. The shrubs then became larger, more vigorous, and established in higher densities.[32]

Although the incidence of natural wildfires was down because of the lack of herbaceous fuel, stockman in those years widely practiced promiscuous fall burning to enhance the growth of herbaceous species for their stock.[33] Such burning favored spring-flowering shrubs such as the *Purshia* species, which were and still are considered good forage for livestock.

If the extensive antelope bitterbrush stands became established in 1890–1900 because the native perennial bunchgrasses were destroyed, then it apparently follows that they reaped the benefits of nutrient-rich soils dominated by perennial grasses for centuries before the introduction of domestic livestock. If this is true, we are now trying to restore antelope bitterbrush on sites with *different* soil parameters than existed when the original stands were established.

The management of *Purshia* species has focused on their importance as browse plants for both native and domestic herbivores, but more particularly on the importance of antelope bitterbrush to mule deer. Since antelope bitterbrush was first recognized as important forage for mule deer, its management has circled around livestock grazing. Livestock-wildlife conflicts resulted in the implementation of many research programs that provided usable data and useful observations on the management of antelope bitterbrush, but the focus of this livestock-wildlife issue may also have deterred proper management of antelope bitterbrush stands. For example, in the 1950s A. Starker Leopold suggested that deliberate manipulation of the vegetation was necessary to maintain high carrying capacities for mule deer.[34] Then, as now, wildlife managers frowned on the practices of vegetation manipulation, which were viewed as favoring livestock over deer. Manipulation practices such as prescribed burning set back succession, which immediately encourages the herbaceous forage favored by livestock operators. *If* properly managed (i.e., regarding stocking rates and season of use), however, these ranges can become succulent and diverse plant communities made up of different successional stages that will provide excellent wildlife habitat for years.

Foliage production of antelope bitterbrush is reported to peak at about 60 years of age.[35] If this is so, many antelope bitterbrush stands today are well past their prime. We aged antelope bitterbrush stands in three locations in northeastern California and northwestern Nevada and found only one stand that had an average age less than 60 years, on private property that had been cleared by equipment in the early 1960s. The antelope bitterbrush shrubs there averaged 33 years of age and were quite vigorous. Plants on the worst-looking site (in terms of plant vigor) averaged 98 years of age, well past their peak. This site is right in the middle of a major migration corridor for the Lassen interstate mule deer herd and has long been a bone of contention between those who want to manage it for livestock and those who think it should be managed for mule deer.

Fig. 12.6. The nature of the problem: one lone antelope bitterbrush plant, distorted by excessive browsing, in a stand of big sagebrush with juniper encroaching in the background. No antelope bitterbrush seedling recruitment is apparent.

Purshia species, especially antelope bitterbrush, have received a lot of attention because of their importance as a browse. Yet *Purshia* management may be just as inefficient today as it was 50 years ago when game managers first grew concerned about declining browse species. Although the *Purshia* species appear to have increased following dramatic events at the turn of the century, the reasons for the increase are still not well understood. When you sit down on the range and carefully consider all the different events that affect rangelands, both natural and artificial, and the diverse characteristics that make up range plant communities, the reasons why managing *Purshia* species is so difficult and complex are obvious.

Reality in antelope bitterbrush management is shown in Figure 12.6, a single plant that has been distorted by excessive browsing. Heavy competition from cheatgrass, big sagebrush, and invading juniper trees prevents seedling recruitment. If restoration of antelope bitterbrush is required for mule deer habitat maintenance, a major weed control and seeding regime must be established. There are no simple answers for sites in this condition. Removing domestic livestock grazing accomplishes nothing. Passive management will lead to burning in a wildfire and complete loss of the antelope bitterbrush. Mule deer habitat managers face very difficult challenges.

Notes

Chapter One. The Wild and Bitter Roses

1. D. I. Axelrod, *Contributions to Paleontology*, VI: *Evolution of Desert Vegetation in Western North America* (Carnegie Inst. Wash. Publ. 590, 1950), pp. 215–306.

2. The terminology used for the two bitterbrushes and cliffrose follows *The Jepson Manual: Higher Plants of California*, ed. J. C. Hickman (Berkeley: Univ. Calif. Press, 1993). We discuss later the practical significance of this classification.

3. Willis Linn Jepson, *Flora of California*, vol. 2 (San Francisco: Calif. School Book Depository, 1936). Jepson quoted from Pursh, "prairies of the Rocky Mountains," as the original collection location. Perhaps Pursh had more than one specimen from the Lewis and Clark Expedition.

4. H. Stansbury, *Exploration and Survey of the Valley of the Great Salt Lake of Utah, Including a Reconnaissance of a New Route through the Rocky Mountains* (Philadelphia: Lippincott, Grambo, 1852).

5. A. L. Hormay, "Bitterbrush in California" (Res. Note 34, USDA, Forest Serv., Berkeley, Calif., 1943). This is one of the classic papers on the ecology of antelope bitterbrush.

6. D. L. Koehler and D. M. Smith, "Hybridization between *Cowania mexicana* var. *stansburiana* and *Purshia glandulosa* (Rosaceae)," *Madroño* 28 (1981): 13–25.

7. Ibid.

8. E. D. McArthur, H. C. Stutz, and S. C. Sanderson, "Taxonomy, distribution, and cytogenetics of *Purshia, Cowania,* and *Fallugia* (Rosoideae, Rosaceae)," pp. 4–24 in *Proceedings of the Symposium on Research and Management of Bitterbrush and Cliffrose in Western North America,* ed. A. R. Tiedemann and K. L. Johnson (Gen. Tech. Rep. 152, USDA, Forest Serv., Ogden, Utah, 1983).

9. Information on rose family from Barbara Ertter and Victor H. Wilken, "Rosaceae," pp. 942–943 in Hickman 1993. The classic reference for the rose family is K. R. Robertson, "The genera of Rosaceae in the southeastern United States," *J. Arnold Arboretum* 55 (1974): 303–332, 344–401, 611–662.

10. Peter H. Raven and Daniel I. Axelrod, "Angiosperm biogeography and past continental movements," *Ann. Mo. Bot. Garden* 61 (1974): 539–673. This is the classic monograph on angiosperm biography as interpolated with the modern theories of continental drift.

11. A. Cronquist, *The Evolution and Classification of Flowering Plants* (Boston: Houghton Mifflin, 1968).

12. Raven and Axelrod, "Angiosperm biogeography."

13. Ibid.

14. McArthur et al., "Taxonomy, distribution, and cytogenetics of *Purshia.*" Dr. Stutz gave an inspiring presentation of this paper as the keynote address at this symposium.

15. R. W. Chaney, *Introduction to Pliocene Floras of California and Oregon* (Carnegie Inst. Wash. Publ. 553, 1944), pp. 1–19; R. W. Chaney and D. I. Axelrod, *Miocene Floras of the Columbia Basin* (Carnegie Inst. Wash. Publ. 617, 1959), pp. 1–237; Axelrod, *Evolution of Desert Vegetation.*

16. D. I. Axelrod, *Mio-Pliocene Floras from West-Central Nevada* (Univ. Calif. Publ. Geol. Sci. 33, 1956), pp. 1–316.

17. The animal names were taken as representative fauna from J. R. MacDonald, "A new Clarendonian mammalian fauna from the Truckee formation of western Nevada," *J. Paleontol.* 30 (1956): 186–202.

18. McArthur et al., "Taxonomy, distribution, and cytogenetics of *Purshia.*" The authors based these conclusions on extensive review of published sources and personal communication with D. I. Axelrod.

19. T. Brandegee, "Flora of the Providence Mountains," *Zoe* 5 (1903): 148–153.

20. G. Ledyard Stebbins, "The role of hybridization in evolution," *Proc. Am. Philos. Soc.* 103 (1959): 231–251.

21. E. Anderson, *Introgressive Hybridization* (New York: Wiley and Sons, 1949).

22. Stebbins, "The role of hybridization," p. 240.

23. H. C. Stutz and L. K. Thomas, "Hybridization and introgression in *Cowania* and *Purshia,*" *Evolution* 18 (1963): 183–195; L. K. Thomas Jr., "Introgression in *Purshia tridentata* (Pursh) DC and *Cowania stansburiana* Torr." (M.S. thesis, Brigham Young Univ., Provo, Utah, 1957).

24. McArthur et al., "Taxonomy, distribution, and cytogenetics of *Purshia.*"

25. Ibid.

26. Stutz and Thomas, "Hybridization and introgression in *Cowania* and *Purshia.*"

27. Thomas J. Rosattii, "*Purshia,*" pp. 970–971 in Hickman 1993.

28. R. Watkins, "Apple and pear," pp. 247–250 in *Evolution of Crop Plants,* ed. N. W. Simmonds (London: Longman, 1976).

29. Jean Marie Alderfer, "A taxonomic study of bitterbrush (*Purshia tridentata* [Pursh] DC) in Oregon" (M.S. thesis, Ore. State Univ., Corvallis, 1977). This thesis provides details of significant variation within *Purshia tridentata.*

30. Nancy Shaw and S. B. Monsen, "'Lassen' antelope bitterbrush," pp. 364–372 in *Proceedings of the Wildland Shrub and Arid Land Restoration Symposium,* ed. B. A. Roundy, E. D. McArthur, J. S. Haley, and D. K. Mann (Gen. Tech. Rep. 315, USDA, Forest Serv., Odgen, Utah, 1995).

Chapter Two. Hunters, Herdsmen, and Brush

1. Arthur W. Sampson, *Native American Forage Plants* (New York: Wiley and Sons, 1924). At the time this book was published Sampson was associate professor of range

management and forest ecology at the University of California, Berkeley. Before that he was plant ecologist and director of the Great Basin Experiment Station, USDA, Forest Serv., in the Intermountain Region. All Sampson quotations are from this source.

2. *Forage Conditions and Problems in Eastern Washington, Eastern Oregon, Northeastern California, and Northwestern Nevada* (Bull. 15, USDA, Bur. Plant Industry, Washington, D.C., 1902).

3. R. Kent Bearrie, *Plants Used by Sheep on the Mica Mountains Summer Range* (Bull. 113, State Coll. Wash., Pullman, 1913).

4. C. L. Forsling and Earle V. Storm, *The Utilization of Browse Forage as Summer Range for Cattle in Southwestern Utah* (Circ. 62, USDA, Forest Serv., Washington, D.C., 1929).

5. W. N. Sparhauk, *Effect of Grazing upon Western Yellow Pine Reproduction in Central Idaho* (Bull. 738, USDA, Forest Serv., Washington, D.C., 1918).

6. A. E. Aldous and H. L. Shantz, "Types of vegetation in the semiarid portion of the United States and their economic significance," *J. Agric. Res.* 28 (1924): 99–127.

7. William A. Dayton, *Important Western Browse Plants* (Misc. Publ. 101, USDA, Forest Serv., Washington, D.C., 1931).

8. *Range Plant Handbook* (USDA, Forest Serv., Washington, D.C., 1937).

9. J. S. Dixon, "A study of the life history and food habits of mule deer in California," *Calif. Fish and Game* 30.3 (1934): 181–282.

10. The information on the subspecies of mule deer is from the standard text on the subject: W. P. Taylor, ed., *The Deer of North America* (Harrisburg, Pa.: Stackpole, 1969), specifically I. M. Cowan's chapter: "What and where are the mule and black-tailed deer?" pp. 338–362.

11. J. S. Dixon, "What do deer eat?" *Am. Forest and Forest Life* 34.4 (1928): 143–145.

12. E. P. Cliff, "Relation between elk and mule deer in the Blue Mountains of Oregon," *Trans. North Am. Wildl. Conf.* 4 (1939): 560–569.

13. Ibid.

14. C. M. Aldous, "A winter study of mule deer in Nevada," *J. Wildl. Manage.* 9 (1945): 145–151.

15. Oliver T. Edwards, "Survey of winter deer ranges in Malheur National Forest, Oregon," *J. Wildl. Manage.* 6 (1944): 210–220.

16. A. S. Einarsen, "Nine-year observations of deer problem area," *Trans. North Am. Wildl. Conf.* 12 (1947): 193–203.

17. A. S. Leopold, "Deer in relation to plant succession," *Trans. North Am. Wildl. Conf.* 15 (1959): 571–578.

18. T. A. Reynolds Jr., *The Mule Deer: Its History, Life History and Management in Utah* (Bull. 60-4, Utah Dep. Fish and Game, Salt Lake City, 1960).

19. R. J. Costley, "Crippling losses among mule deer in Utah," *Trans. North Am. Wildl. Conf.* 13 (1948): 451–458.

20. Reynolds, *The Mule Deer.*

21. Ross Leonard, *Status and Trends of Big Game in the United States* (Salt Lake City: Utah Fish and Game Dep., 1946).

22. W. M. Longhurst, A. S. Leopold, and R. F. Dasmann, *A Survey of California Deer Herds: Their Ranges and Management Problems* (Game Bull. 6, Calif. Dep. Fish and Game, Sacramento, 1952).

23. C. D. Clements and J. A. Young, "Improved rangeland health and mule deer habitat," *J. Range Manage.* 50 (1997): 129–136.

Chapter Three. Bitterbrush Plant Communities

1. An early example of Daubenmire's work is "Forest vegetation of northern Idaho and adjacent Washington and its bearing on concepts of vegetation classification," *Ecol. Monogr.* 22 (1952): 301–330. *Steppe Vegetation of Washington* (Tech. Bull. 62, Wash. Agric. Exp. Stn., Pullman, 1970) summarizes much of his synecology research. A comprehensive expression of his concepts of community and large plant formation classification can found in *Plant Geography with Special Reference to North America* (New York: Academic Press, 1978).

2. See E. de Wildeman, *Actes des III. Congress International de Botanique,* vol. 1 (Jena: Gustav Fischer, 1910).

3. P. T. Tueller, D. H. Heinze, and R. E. Eckert Jr., "A tentative list of existing Nevada plant communities (a second approximation)" (Coll. Agric., Univ. Nev., Reno, n.d.).

4. J. F. Franklin and C. T. Dyrness, *Natural Vegetation of Oregon and Washington* (Corvallis: Ore. State Univ. Press, 1984; reprint of the bulletin published in 1974 by the USDA, Forest Serv., Portland, Ore).

5. Richard S. Driscoll, "Characteristics of some ecosystems in the juniper zone in central Oregon," *J. Range Manage.* 15 (1962): 347; "A relict area in the central Oregon juniper zone," *Ecology* 45 (1964): 345–353; and "Vegetation-soil units in the central Oregon juniper zone" (Res. Pap. 19, USDA, Forest Serv., Portland, Ore., 1964).

6. Frank Vasek and Robert F. Thorne, "Transmontane coniferous vegetation," pp. 797–832 *in Terrestrial Vegetation of California,* ed. Michael Barbour and Jack Major (New York: Wiley and Sons, 1977).

7. Personal communication from Robin Tausch, who visited the stand to collect western juniper, October 11, 1993. Eamor Nord included this population on a map in a research note; see E. C. Nord, "Bitterbrush ecology—some recent findings" (Res. Note 148, USDA, Forest Serv., Berkeley, Calif., 1959).

8. James A. Young, Raymond A. Evans, and Jack Major, "Sagebrush steppe," pp. 763–793 in Barbour and Major 1977.

9. Nord, "Bitterbrush ecology."

10. Personal communication from Robin Tausch, October 11, 1993.

11. Ibid.

12. Robert J. Sherman, "Spatial and chronological patterns of *Purshia tridentata* as influenced by *Pinus ponderosa* overstory" (M.S. thesis, Ore. State Univ., Corvallis, 1966).

13. "An analysis of montane forest vegetation on the east flank of the central Oregon Cascades" (Ph.D. diss., Ore. State Univ., Corvallis, 1964).

14. Franklin and Dyrness, *Natural Vegetation of Oregon and Washington.*

15. Alastair McLean, "Plant communities of the Similkamean Valley, British Columbia, and their relationship to soils," *Ecol. Monogr.* 40 (1970): 403–424.

16. Franklin and Dryness, *Natural Vegetation of Oregon and Washington.*

17. C. T. Dyrness and C. T. Youngberg, "Soil-vegetation relationships within the ponderosa pine type in the central Oregon pumice region," *Ecology* 47 (1966): 122–138.

18. J. Edward Dealy, "Habitat characteristics of the Silver Lake mule deer range" (Res. Pap. 125, USDA, Forest Serv., Portland, Ore., 1971).

19. Frederick C. Hall, "Vegetation-soil relations as a basis for recourse management on the Ochoco National Forest of central Oregon" (Ph.D. diss., Ore. State Univ., Corvallis, 1967).

20. Robert Steele, Robert D. Pfister, and Jay A. Kittams, "Forest habit types of central Idaho" (Gen. Tech. Rep. 114, USDA, Forest Serv., Ogden, Utah, 1978).

21. Leslie W. Gysel, "An ecological study of the winter range of elk and mule deer in the Rocky Mountain National Park," *J. Forest.* 58 (1960): 696–703.

22. J. M. Peek, F. D. Johnson, and N. N. Perce, "Successional trends in a ponderosa pine/bitterbrush community related to grazing by livestock, wildlife, and to fire," *J. Range Manage.* 31 (1978): 49–53.

23. Philip W. Rundel, David J. Parsons, and Donald T. Gordon, "Montane and subalpine vegetation of the Sierra Nevada and Cascade Range," pp. 559–599 in Barbour and Major 1977.

24. L. A. Volland, "Phytosociology of the ponderosa pine type on pumice soils in the Upper Williamson River Basin, Klamath County, OR" (master's thesis, Ore. State Univ., Corvallis, 1963); Volland, "A multivariate classification of lodgepole pine types in central Oregon with implications of natural resource management" (Ph.D. diss., Colo. State Univ., Fort Collins, 1974).

25. C. T. Youngberg and W. G. Dahms, "Productivity indices for lodgepole pine on pumice soils," *J. Forest.* 68 (1970): 90–94.

26. Paul J. Edgerton, Burt R. McConnell, and Justin G. Smith, "Initial response of bitterbrush to disturbance by logging and slash disposal in a lodgepole pine forest," *J. Range Manage.* 28 (1975): 112–114.

27. David A. Perry and James E. Latan, "Regeneration and early growth of strip clearcuts in a lodgepole pine/bitterbrush habitat type" (Res. Note 238, USDA, Forest Serv., Ogden, Utah, 1977).

28. J. Edward Dealy, "Habitat and characteristics of the Silver Lake mule deer range" (Res. Pap. 125, USDA, Forest Serv., Portland, Ore., 1971).

29. Walter Van-Gale Johnson, "Taxonomy and ecology of the vascular plants of Black Butte, Oregon" (master's thesis, Ore. State Univ., Corvallis, 1959).

30. W. H. Rickard and R. H. Sauer, "Primary production and canopy cover in a bitterbrush-cheatgrass community," *Northwest Sci.* 56 (1982): 250–255.

31. Dixie R. Smith, "Description and response to elk use of two mesic grassland and shrub communities in the Jackson Hole region of Wyoming," *Northwest Sci.* 34 (1960): 25–36.

32. Richard S. Driscoll, "Vegetation-soil units in the central Oregon juniper zone" (Res. Pap. 19, USDA, Forest Serv., Portland, Ore., 1964).

33. Richard S. Driscoll, "A relict area in the central Oregon juniper zone," *Ecology* 45 (1964): 345–353.

34. N. E. West, "Basic synecological relationships of sagebrush dominated lands in the Great Basin and Colorado Plateau," pp. 12–17 in *The Sagebrush Ecosystem* (Logan: Utah State Univ., 1979).

35. Franklin and Dyrness, *Natural Vegetation of Oregon and Washington.*

36. Richard E. Eckert Jr., "Vegetation-soil relationships in some *Artemisia* types in northern Harney and Lake Counties, Oregon" (Ph.D. diss., Ore. State Univ., Corvallis, 1957); Paul T. Tueller, "Plant succession on two *Artemisia* habitat types in southeastern Oregon" (Ph.D. diss., Ore. State Univ., Corvallis, 1962).

37. Ronald K. Tew, "Bitterbrush distribution and habitat classification on the Boise National Forest," pp. 32–36 in *Proceedings of the Symposium on Research and Management of Bitterbrush and Cliffrose in Western North America,* ed. A. R. Tiedemann and K. L. Johnson (Gen. Tech. Rep. 152, USDA, Forest Serv., Ogden, Utah, 1983).

38. J. A. Young, J. R. Wight, and J. E. Mowbray, "Field stratification of antelope bitterbrush seeds," *J. Range Manage.* 46 (1993): 325–330.

39. Wilbert H. Blackburn, R. E. Eckert Jr., and P. T. Tueller, "Vegetation and soils of the Rock Springs watershed" (R 83, Agric. Exp. Stn., Univ. Nev., Reno, 1971).

40. P. T. Tueller and R. E. Eckert Jr., "Big sagebrush (*Artemisia tridentata* subsp. *vaseyana*) and longleaf snowberry (*Symphoricarpus oreophilus*) plant associations in northeastern Nevada," *Great Basin Nat.* 47 (1987): 117–131.

41. Robert H. Berg, "An evaluation of selected Nevada deer ranges: Condition, forage potential, and deer livestock competition" (master's thesis, Univ. Nev., Reno, 1966).

42. N. E. West, K. H. Rea, and Robin Tausch, "Basic synecological relationships in pinyon-juniper woodlands," pp. 41–53 in *The Pinyon Juniper Ecosystem,* ed. G. F. Gifford and F. E. Busby (Logan: Utah State Univ., 1975).

43. Berg, "Evaluation of selected Nevada deer ranges."

44. A. S. Leopold, "Big game management," in *Survey of Fish and Game Problems in Nevada* (Bull. 36, Nev. Legislative Council Bur., Carson City, 1959). Also see Wilbert H. Blackburn and P. T. Tueller, "Pinyon and juniper invasion in black sagebrush communities in east-central Nevada," *Ecology* 51 (1970): 841–848.

45. J. A. Young and J. D. Budy, "Historical use of Nevada's pinyon-juniper woodlands," *J. Forest. Hist.* 23 (1979): 113–121.

46. P. T. Tueller, A. D. Brunner, and J. Barry Davis, "Ecology of Hot Creek Valley" (R 89, Agric. Exp. Stn., Univ. Nev., Reno, 1972).

47. B. J. Albee, *Atlas of the Vascular Plants of Utah* (Occas. Pap. 8, Utah Mus. Nat. Hist., Salt Lake City, 1988).

48. R. D. Pieper and G. A. Lymbery, "Influence of topographic features on pinyon/juniper vegetation in south-central New Mexico," pp. 53–57 in *Proceedings of the Pinyon/Juniper Conference,* comp. R. L. Everett (Gen. Tech. Rep. 215, USDA, Forest Serv., Ogden, Utah, 1987).

49. W. H. Moir and J. O. Carleton, "Classification of pinyon/juniper sites on national forest in the Southwest," pp. 216–226 in Everett 1987.

50. Nord, "Bitterbrush ecology."

51. K. P. Price and J. D. Brotherson, "Habitat and community relationships of cliffrose (*Cowania mexicana* var. *stansburiana*) in central Utah," *Great Basin Nat.* 47 (1987): 132–151.

Chapter Four. Ecophysiology of Purshia

1. Eamor C. Nord, "Autecology of bitterbrush in California," *Ecol. Monogr.* 35 (1965): 307–334.

2. August L. Hormay, "Bitterbrush in California" (Res. Note 34, USDA, Forest Serv., Berkeley, Calif., 1943).

3. The Idaho data Nord used came from J. P. Blaisdell, *Seasonal Development and Yield of Native Plants on the Upper Snake River Plains and Their Relation to Certain Climatic Factors* (Tech. Bull. 1190, USDA, Washington, D.C., 1958).

4. Nancy L. Shaw and Stephen B. Monsen, "Phenology and growth habits of nine antelope bitterbrush, desert bitterbrush, Stansbury cliffrose, and Apache-plume accessions," pp. 55–69 in *Proceedings of the Symposium on Research and Management of Bitterbrush and Cliffrose in Western North America,* ed. A. R. Tiedemann and K. L. Johnson (Gen. Tech. Rep. 152, USDA, Forest Serv., Odgen, Utah, 1983).

5. Jean Marie Alderfer, "A taxonomic study of bitterbrush (*Purshia tridentata* [Pursh] DC.) in Oregon" (M.S. thesis, Ore. State Univ., Corvallis, 1976).

6. A. C. Blauer, A. P. Plummer, E. D. McArthur, R. Stevens, and B. C. Giunta, "Characteristics and hybridization of important intermountain shrubs. I. Rose family" (Res. Pap. 169, USDA, Forest Serv., Odgen, Utah, 1975).

7. Ibid.

8. Ibid.

9. See, for example, T. J. Rosatti, "*Purshia,*" pp. 970–971 in *The Jepson Manual: Higher Plants of California,* ed. J. C. Hickman (Berkeley: Univ. Calif. Press, 1993).

10. Nord, "Autecology of bitterbrush."

11. Frank W. Stanton, "Autecological studies of bitterbrush (*Purshia tridentata* [Pursh] DC.)" (Ph.D. diss., Ore. State Univ., Corvallis, 1959).

12. Blaisdell, *Seasonal Development.*

13. E. C. McCarty and R. Price, *Growth and Carbohydrate Content of Important Mountain Forage Plants in Central Utah as Affected by Clipping and Grazing* (Tech. Bull. 818, USDA, Washington, D.C., 1942).

14. B. R. McConnell and G. A. Garrison, "Seasonal variation of available carbohydrates in bitterbrush," *J. Wildl. Manage.* 30 (1996): 168–172.

15. M. Buwai and M. J. Trlica, "Multiple defoliation effects on herbage yield, vigor, and total nonstructural carbohydrates of five range species," *J. Range Manage.* 30 (1977): 164–171.

16. M. M. Caldwell, J. H. Richards, D. A. Johnson, R. S. Nowak, and R. S. Dzurec, "Coping with herbivory: Photosynthetic capacity and resource allocation in two semiarid *Agropyron* bunchgrasses," *Oecologica* 50 (1981): 14–24.

17. Blauer et al., "Characteristics of Intermountain shrubs."

18. Nord, "Autecology of bitterbrush."

19. Alderfer, "Taxonomic study of bitterbrush."

20. Shaw and Monsen, "Phenology." The authors suggest that Nord made the same observation, but Nord's data suggest that the progeny of intermediate and semiprostrate plants were actually segregating for these traits.

21. J. P. Blaisdell and W. F. Mueggler, "Sprouting of bitterbrush (*Purshia tridentata*) following burning or top removal," *Ecology* 37(1956): 365–369.

22. R. G. Clark, C. M. Britton, and F. A. Sneva, "Mortality of bitterbrush after burning and clipping in eastern Oregon," *J. Range Manage.* 35 (1982): 711–714.

23. Eamor C. Nord, "Was this a prize bitterbrush?" *J. Range Manage.* 15 (1962): 82–83.

24. Richard S. Driscoll, "A large bitterbrush," *J. Range Manage.* 16 (1963): 82–83.

25. Blauer et al., "Characteristics of Intermountain shrubs."

26. J. N. Davis, "Performance comparison among populations of bitterbrush, cliffrose, and bitterbrush-cliffrose crosses on study sites throughout Utah," pp. 38–44 in Tiedemann and Johnson 1983.

27. P. J. Edgerton, J. M. Geist, and W. G. Williams, "Survival and growth of Apache-plume, Stansbury cliffrose, and selected sources of antelope bitterbrush in northeastern Oregon," pp. 45–54 in Tiedemann and Johnson 1983.

28. C. J. Bilbrough and J. H. Richards, "Branch architecture of sagebrush and bitterbrush: Use of branch complex to describe and compare patterns of growth," *Can. J. Bot.* 69 (1991): 1288–1295.

29. Stanton, "Autecological studies of bitterbrush."

30. Ibid.

31. Robert R. Kindschy, "Effect of precipitation variance on annual growth of 14 browse shrubs in southeastern Oregon," *J. Range Manage.* 35 (1982): 265–266.

32. Richard L. Hubbard and H. Reed Sanderson, "Herbage production and carrying capacity of bitterbrush" (Res. Note 157, USDA, Forest Serv., Berkeley, Calif., 1950).

33. L. Wandera, J. H. Richards, and R. J. Mueller, "The relationships between relative growth rate, meristematic potential and compensatory growth of semiarid-land shrubs," *Oecologia* 90 (1992): 391–398.

34. G. A. Garrison, "Effects of clipping on some range shrubs," *J. Range Manage.* 6 (1953): 309–317.

35. P. Train, J. R. Henrichs, and W. A. Archer, "Medicinal uses of plants by Indian tribes of Nevada," in *Contributions toward a Flora of Nevada* (No. 33, Works Projects Admin. Nev., Washington, D.C., 1941).

36. Henry Trimble, "Bitterbrush," *Am. J. Pharmacy* 64 (1892): 69; Charles V. Netz, C. H. Rogers, and G. L. Jenkins, "A phytochemical and histological study of *Purshia tridentata*," *J. Am. Pharmacy Assoc.* 29 (1940): 480–485.

Chapter Five. Purshia *Seed Physiology*

1. August L. Hormay, "Bitterbrush in California" (Res. Note 34, USDA, Forest Serv., Berkeley, Calif., 1943).

2. In a paper published in 1958, Richard L. Hubbard suggested that the first person to discover that antelope bitterbrush seeds could be made to germinate by moist prechilling was N. T. Mirov, who, along with Charles J. Kraebel, compiled a volume on propagating seeds of California wildland species for use in conservation plantings. It was published as Forestry Publication 5 of the Civilian Conservation Corps, Berkeley, California, in 1939. See R. L. Hubbard, "Germination of thiourea-treated bitterbrush seed in the field" (Forest Res. Note 138, USDA, Forest Serv., Berkeley, Calif., 1958).

3. E. A. Nord, "Quick testing bitterbrush seed viability," *J. Range Manage.* 9 (1956): 193–194.

4. James A. Young and Raymond A. Evans, "Stratification of bitterbrush seeds," *J. Range Manage.* 29 (1976): 421–425.

5. Eamor C. Nord, "Autecology of bitterbrush in California," *Ecol. Monogr.* 35 (1965): 307–334.

6. James A. Young, J. Ross Wight, and J. E. Mowbray, "Field stratification of antelope bitterbrush seeds," *J. Range Manage.* 46 (1993): 325–330.

7. Young and Evans, "Stratification of bitterbrush seeds."

8. D. Cone, "L'inhibition de germination des graines de Pommier (*Pirus malus* L.) non-dormantes. Rôle possible des phenols tegumentaltes," *Ann. Sci. Nat. Bot and Biol. Veg.* 8 (1967): 371–478.

9. D. T. Booth, "Bitterbrush seed dormancy and seedling vigor" [abstract], Society for Range Management, annual meeting, Omaha, Neb., 1999, 52:6.

10. G. A. Harris, "Some competitive relationships between *Agropyron spicatum* and *Bromus tectorum*," *Ecol. Monogr.* 37 (1967): 89–111.

11. Robert Fay Wagle, "Early growth in bitterbrush and its relation to environment" (Ph.D. diss., Univ. Calif., Berkeley, 1958).

12. *Woody Plant Seed Manual* (Handb. 654, USDA, Forest Serv., Washington, D.C., 1948).

13. R. A. Peterson, "Comparative effects of seed treatments upon seedling emergence of seven browse species," *Ecology* 34 (1963): 778–785.

14. B. O. Pearson, "Bitterbrush seed dormancy broken with thiourea," *J. Range Manage.* 10 (1957): 41–42.

15. See, for example, C. G. Deuber, "Chemical treatments to shorten the rest period of red and black oak acorns," *J. Forest.* 30 (1932): 674–679.

16. A. M. Mayer and A. Poljakoff-Mayber, *The Germination of Seeds*, 4th ed. (Oxford: Pergamon Press, 1989). This is the classic review source for seed physiology, but in this case the authors also did much of the original research on the subject.

17. Hubbard, "Germination of thiourea-treated bitterbrush seed."

18. Young et al., "Field stratification."

19. E. C. Nord and G. R. Van Atta, "Saponin—a seed germination inhibitor," *Forest Sci.* 6 (1960): 350–353.

20. David L. Dreyer and Eugene K. Trousdale, "Cuccurbitacins in *Purshia tridentata*," *Phytochemistry* 17 (1978): 325–326.

21. See, for example, E. R. Brown and C. F. Martinsen, *Browse Planting for Big Game* (Biol. Bull. 12, Wash. State Game Dep., Olympia, 1959).

22. Donald L. Neal and H. Reed Sanderson, "Thiourea solution temperature and bitterbrush germination and seedling growth," *J. Range Manage.* 28 (1975): 421–423.

23. L. W. Harper, "The use of thiourea for laboratory germination of antelope bitterbrush seed," *Proc. Assoc. Off. Seed Anal.* 60 (1970): 127–131.

24. Richard L. Everett and Richard O. Meeuwig, "Hydrogen peroxide and thiourea treatment of bitterbrush seed" (Res. Note 196, USDA, Forest Serv., Odgen, Utah, 1975).

25. R. S. Smith, R. F. Scharpf, and E. R. Schneegas, "Frost injury to bitterbrush in eastern California" (Res. Note 82, USDA, Forest Serv., Berkeley, Calif., 1965).

26. B. R. McConnell, "Effect of gibberellic acid and cold treatments on the germination of bitterbrush seed" (Res. Note 187, USDA, Forest Serv., Portland, Ore., 1960).

27. W. Crocker, "Mechanics of dormancy," *Am. J. Bot.* 3 (1916): 99–120.

28. See, for example, J. M. Trappe, "Strong hydrogen peroxide for sterilizing coats of tree seeds and stimulating germination," *J. Forest.* 59 (1961): 828–829.

29. J. W. Riffle and H. W. Springfield, "Hydrogen peroxide increases germination and reduces microflora on seed of several southwestern woody species," *Forest Sci.* 14 (1968): 96–101.

30. James A. Young and Raymond A. Evans, *Germination of Seeds of Antelope Bitterbrush, Desert Bitterbrush, and Cliffrose* (Agric. Res. Results 17, USDA, Agric. Res. Ser., Oakland, Calif., 1981).

31. R. A. Evans, D. E. Palmquist, D. N. Book, and J. A. Young, "Quadratic response surface analysis of seed-germination trials," *Weed Sci.* 30 (1982): 411–416; Debra Palmquist, R. A. Evans, and J. A. Young, "Comparative analysis of temperature-germination response surfaces," pp. 97–104 in *Proceedings of the Symposium on Seed and Seedbed Ecology of Rangeland Plants,* ed. G. W. Frasier and R. A. Evans (USDA, Agric. Res. Ser., Washington, D.C., 1987). The various germination parameters calculated from the germination-temperature profiles are defined in J. A. Young and R. A. Evans, *Temperature Profiles for Germination of Cool Season Range Grasses* (Agric. Res. Results 27, USDA, Agric. Res. Ser., Oakland, Calif., 1982).

32. Pearson, "Bitterbrush seed dormancy broken."

33. Young and Evans, "Stratification of bitterbrush seeds."

34. R. A. Evans, H. R. Holbo, R. E. Eckert Jr., and J. A. Young, "Functional environment of downy brome communities in relation to weed control and revegetation," *Weed Sci.* 18 (1970): 154–162.

Chapter Six. Seeding Purshia *Species*

1. August L. Hormay, "Bitterbrush in California" (Res. Note 34, USDA, Forest Serv., Berkeley, Calif., 1943).

2. Nord first published his bioclimatic equation for predicting the maturity of antelope bitterbrush seeds in E. C. Nord, "Bitterbrush seed harvesting: When, where, and how," *J. Range Manage.* 16 (1963): 258–261. He repeated this material, perhaps with some elaboration, in E. C. Nord, "Autecology of bitterbrush in California," *Ecol. Monogr.* 35 (1965): 307–334. Eamor did not cite the *Journal of Range Management* paper in the later monograph.

3. In a later publication Richard Hubbard provided a photograph of the same tray and reported that it was designed by Hugo Herman of the California Department of Fish and Game. See Richard L. Hubbard, "A guide to bitterbrush seeding in California" (Res. Note 34, USDA, Forest Serv., Berkeley, Calif., 1964).

4. Eamor C. Nord, Edward Schneegas, and Hatch Graham, "Bitterbrush seed collecting—by machine or hand," *J. Range Manage.* 20 (1967): 99–102.

5. Ibid.; B. C. Giunta, R. Stevens, K. R. Jorgensen, and A. P. Plummer, *Antelope Bitterbrush — an Important Wildland Shrub* (Publ. 78–12, Utah Div. Wildl. Res., Salt Lake City, 1978).

6. Richard Steves and Kent R. Jorgensen, "Rangeland species germination through 25 and up to 40 years of warehouse storage," pp. 257–265 in *Proceedings of a Symposium on Ecology and Management of Annual Rangelands,* ed. S. B. Monsen and S. G. Kitchen (Gen. Tech Rep. 313, USDA, Forest Serv., Ogden, Utah, 1994).

7. For a discussion of storage procedures, see J. A. Young and Cheryl G. Young, *Collecting, Processing, and Germinating Seeds of Wildland Plants* (Portland, Ore.: Timber Press, 1985).

8. Nord et al., "Bitterbrush seed collecting."

9. R. B. Ferguson, "Relative germination of spotted and non-spotted bitterbrush seed," *J. Range Manage.* 20 (1967): 330–331.

10. J. A. Young and R. A. Evans, "Population dynamics after wildfires in sagebrush grasslands," *J. Range Manage.* 31 (1967): 283–289.

11. T. A. Phillips, "The status of antelope bitterbrush in the Cassia Mountain area of southern Idaho," *Range Improvement Notes* 15.4 (1970): 1–15.

12. J. A. Young and R. A. Evans, "Germination and establishment of *Salsola* in relation to seedbed environment. Part I. Temperature, afterripening, and moisture relations of *Salsola* as determined by laboratory studies," *Agron. J.* 64 (1972): 214–218.

13. R. L. Hubbard and H. R. Sanderson, "When to plant bitterbrush—spring or fall?" (Tech. Pap. 64, USDA, Forest Serv., Berkeley, Calif., 1961).

14. R. A. Evans, H. R. Holbo, R. E. Eckert Jr., and J. A. Young, "Functional environment of downy brome communities in relation to weed control and revegetation," *Weed Sci.* 18 (1970): 89–97.

15. R. B. Ferguson and J. V. Basile, "Effect of seedling numbers on bitterbrush survival," *J. Range Manage.* 20 (1967): 380–382.

16. Ralph C. Holmgren, "Competition between annuals and young bitterbrush (*Purshia tridentata*) in Idaho," *Ecology* 37 (1956): 370–377.

17. Ralph C. Holmgren and Joseph V. Basile, *Improving Southern Idaho Deer Winter Ranges by Artificial Revegetation* (Wildl. Bull. 3, Idaho Dep. Fish and Game, Boise, 1959).

18. R. B. Ferguson, "Bitterbrush seedling establishment as influenced by soil moisture and soil surface temperature," *J. Range Manage.* 25 (1972): 47–49.

19. R. B. Ferguson, "Growth of single bitterbrush plants vs. multiple group established by direct seeding" (Res. Note 90, USDA, Forest Serv., Ogden, Utah, 1962).

20. J. A. Young and D. McKenzie, "Rangeland drill," *Rangelands* 4 (1982): 108–113. This publication provides a history of the range improvement era in the Intermountain area.

21. R. L. Hubbard, E. C. Nord, and L. L. Brown, "Bitterbrush reseeding . . . a tool for the game range manager" (Misc. Pap. 39, USDA, Forest Serv., Berkeley, Calif., 1959).

22. R. L. Hubbard, "Effect of depth of planting on emergence and survival of bitterbrush seedlings" (Res. Note 113, USDA, Forest Serv., Berkeley, Calif., 1965).

23. J. L. Harper, *Population Biology of Plants* (London: Academic Press, 1977).

24. J. V. Basile and R. C. Holmgren, "Seeding-depth trials with bitterbrush (*Purshia tridentata*) in Idaho" (Res. Pap. 54, USDA, Forest Serv., Ogden, Utah, 1957).

25. Hormay, "Bitterbrush in California," 5.

26. Young and McKenzie, "Rangeland drill."

27. Hubbard et al., "Bitterbrush reseeding."

28. Eamor C. Nord and Bert Knowles, "Rice hulls improve drilling of bitterbrush seed" (Res. Note 134, USDA, Forest Serv., Berkeley, Calif., 1958).

29. Hormay, "Bitterbrush in California," 11.

30. R. L. Hubbard, "The effects of plant competition upon the growth and survival of bitterbrush seedlings" (Res. Note 109, USDA, Forest Serv., Berkeley, Calif., 1956).

31. Memo: Report of bitterbrush seeding on the Paisley Ranger District, Fremont National Forest (USDA, Forest Serv., Lakeview, Ore., 1950).

32. R. L. Hubbard, "A guide to bitterbrush seeding in California" (Res. Note 34, USDA, Forest Serv., Berkeley, Calif., 1964).

33. The study is eloquently summarized in R. L. Hubbard, "The place of browse seeding in game range management," pp. 394–401 in *Trans. 27th North Am. Wildl. Nat. Res. Conf.*, 1962.

34. Phillips, "Status of antelope bitterbrush in the Cassia Mountain area."

35. R. L. Hubbard, P. Zusman, and R. Sanderson, "Bitterbrush stocking and minimum spacing with crested wheatgrass," *Calif. Fish and Game* 48.3 (1962): 203–208. Also see Howard R. Leach, "Food habits of the Great Basin deer herds of California," *Calif. Fish and Game* 42.4 (1956): 243–308.

36. Stephen B. Monsen and Nancy L. Shaw, "Seeding antelope bitterbrush with grasses on south-central Idaho rangelands—a 39 year response," pp. 126–136 in *Proceedings of the Symposium on Research and Management of Bitterbrush and Cliffrose in Western North America*, ed. A. R. Tiedemann and K. L. Johnson (Gen. Tech. Rep. 152, USDA, Forest Serv., Ogden, Utah, 1983).

37. R. M. Hurd, "A range test for species adaptability," *Iowa State Coll. J. Sci.* 22 (1948): 387–394.

38. Holmgren, "Competition between annuals and young bitterbrush."

39. R. L. Piemeisel, "Causes affecting change and rate of change in a vegetation of annuals in Idaho," *Ecology* 32 (1951): 53–72. This summary article was written at the end of Piemeisel's productive career.

40. J. H. Robertson and C. K. Pearse, "Artificial reseeding and the closed community," *Northwest Sci.* 19 (1945): 58–66.

41. J. A. Young, R. A. Evans, and R. E. Eckert Jr., "Population dynamics of downy brome," *Weed Sci.* 17 (1969): 20–26.

42. R. C. Holmgren and J. V. Basile, *Improving Southern Idaho Deer Winter Ranges by Artificial Revegetation* (Bull. 3, Idaho Dep. Fish and Game, Boise, 1959).

43. Ibid., 23.

44. Young et al., "Population dynamics of downy brome."

45. Ibid.

46. B. C. Giunta, D. R. Christensen, and S. B. Monsen, "Interseeding shrubs in cheatgrass with a browse seeder-scalper," *J. Range Manage.* 28 (1975): 398–402.

47. R. B. Ferguson, "Survival and growth of young bitterbrush browsed by deer," *J. Wildl. Manage.* 32 (1968): 769–772.

48. J. Edward Dealy, "Survival and growth of bitterbrush on the Silver Lake deer winter range in central Oregon" (Res. Note 133, USDA, Forest Serv., Portland, Ore., 1970).

49. R. A. Evans, R. E. Eckert Jr., B. L. Kay, and J. A. Young, "Downy brome control by soil-active herbicides for revegetation of rangelands," *Weed Sci.* 17 (1969): 166–169.

50. R. A. Evans and J. A. Young, "Microsite requirements for establishment of alien annual weed species in rangeland communities," *Weed Sci.* 20 (1972): 350–356.

51. M. Dale Christensen, J. A. Young, and R. A. Evans, "Control of annual grasses and revegetation in ponderosa pine woodlands," *J. Range Manage.* 27 (1974): 143–145.

Chapter Seven. Granivore Relations

1. E. E. Horn and H. S. Fitch, "Interrelations of rodents and other wildlife on the range," *Calif. Dep. Agric. Bull.* 663 (1942): 96–129.

2. Arthur C. Cole Jr., Pogonomyrmex *Harvester Ants* (Knoxville: Univ. Tenn. Press, 1968).

3. 6. S. B. Vander Wall, *Food Hoarding in Animals* (Chicago: Univ. Chicago Press, 1990).

4. Cole, Pogonomyrmex *Harvester Ants.*

5. William H. Clark and Peter L. Comanor, "The use of western harvester ant, *Pogonomyrmex occidentalis* (Cresson), seed stores by heteromyid rodents" (Occas. Pap. 34, Biol. Soc. Nev., Reno, 1973).

6. Raymond A. Evans, J. A. Young, Greg Cluff, and J. Kent McAdoo, "Dynamics of antelope bitterbrush seed caches," pp. 195–203 in *Proceedings of the Symposium on Research and Management of Bitterbrush and Cliffrose in Western North America,* ed.

A. R. Tiedemann and K. L. Johnson (Gen. Tech. Rep. 152, USDA, Forest Serv., Odgen, Utah, 1983).

7. S. B. Vander Wall, "Dispersal and establishment of antelope bitterbrush by seed caching rodents," *Ecology* 17 (1994): 1911–1926.

8. E. R. Schneegas, "Aspen snag yields record bitterbrush seed cache," *J. Range Manage.* 18 (1965): 34–35.

9. Vander Wall, "Dispersal and establishment of antelope bitterbrush by seed caching rodents."

10. R. L. Lassen, C. M. Ferrel, and H. R. Leach, "Food habits, productivity, and condition of the Doyle mule deer herd," *Calif. Fish and Game* 38 (1952): 211–224; T. A. Phillips, "The status of antelope bitterbrush in the Cassia Mountain area of southern Idaho" (Res. Note 15, USDA, Forest Serv., Ogden, Utah, 1970); A. W. Adams, *A Brief History of Juniper and Shrub Populations in Southern Oregon* (Res. Rep. 6, Ore. State Wildl. Comm., Corvallis, 1975); C. D. Clements and J. A. Young, "A viewpoint: Improved rangeland health and mule deer habitat," *J. Range Manage.* 50 (1997): 139–145.

11. Vander Wall, "Dispersal and establishment of antelope bitterbrush by seed caching rodents."

12. S. H. Jenkins and R. A. Peters, "Spatial patterns of food storage by Merriam's kangaroo rats," *Behav. Ecol.* 3 (1992): 60–65.

13. S. R. Morton, R. Hinds, and R. E. MacMillen, "Cheek pouch capacity in heteromyid rodents," *Oecologia* 46 (1980): 143–146.

14. M. I. Kelrick and J. A. MacMahon, "Nutritional and physical attributes of seeds of common sagebrush-steppe plants: Some implications for ecological theory and management," *J. Range Manage.* 38 (1985): 65–69; M. I. Kelrick, J. A. MacMahon, R. R. Parmenter, and D. V. Sisson, "Native seed preference of shrub-steppe rodents, birds, and ants: The relationships of seed attributes and seed use," *Oecologia* 68 (1986): 327–337.

15. C. L. Franks, "The influence of moisture content on seed selection by kangaroo rats," *J. Mammal.* 69 (1988): 353–357.

16. National Research Council, *Nutrient Requirements of Laboratory Animals* (Natl. Res. Council Publ. 10, Washington, D.C., 1978).

17. D. B. Webster and M. Webster, "Adaptive value of hearing and vision in kangaroo rat predator avoidance," *Brain Behav. Evol.* 4 (1971): 310–322.

18. S. H. Jenkins and R. Ascanio, "A potential nutritional basis for resource partitioning by desert rodents," *Am. Midl. Nat.* 130 (1993): 164–172.

19. D. A. Spencer, "Rodents and direct seeding," *J. Forest.* 52 (1954): 824–826.

20. Ibid.

21. R. L. Casebeer, "The use of tetramine in bitterbrush revegetation," *J. Forest.* 52 (1954): 829–830.

22. R. L. Everett, R. O. Meeuwig, and R. Stevens, "Deer mouse preference for seed of commonly planted species, indigenous weed seed, and sacrifice foods," *J. Range Manage.* 31 (1978): 70–73.

23. Evans et al., "Dynamics of antelope bitterbrush seed caches."

24. R. L. Sherman and W. W. Chilcote, "Spatial and chronological patterns of *Purshia tridentata* as influenced by *Pinus ponderosa*," *Ecology* 53 (1972): 294–298.

25. C. D. Clements and J. A. Young, "Influence of rodent predation on antelope bitterbrush seedlings," *J. Range Manage.* 49 (1996): 31–34.

26. R. L. Hubbard and S. McKeever, "Meadow mouse girdling—another cause of death of reseeded bitterbrush plants," *Ecology* 42 (1962): 198.

Chapter Eight. Ruminant Nutrition

1. The following discussion of nutrition as it relates to mule deer and *Purshia* species follows the excellent review paper written by Donald R. Dietz: "Nutritive value of shrubs," pp. 289–302 in *Symposium on Wildland Shrubs — Their Biology and Utilization* (Gen. Tech. Rep. 1, USDA, Forest Serv., Ogden, Utah, 1972). The discussion is also based on works by F. B. Morrison; e.g., *Feeds and Feeding* (Ithaca, N.Y.: Morrison, 1943). For much of the twentieth century this has been the source book for animal nutrition.

2. See, for example, Henry G. Knight, Frank E. Hepner, and Aven Nelson, *Wyoming Forage Plants and Their Chemical Composition* (Bull. 4, Wyo. Agric. Exp., Univ. Wyo., Laramie, 1911).

3. See, for example, A. A. Nichol, *Experimental Feeding of Deer* (Tech. Bull. 75, Agric. Exp. Stn., Univ. Ariz., Tucson, 1938). Cliffrose was among the browse species used in these trials.

4. L. A. Stoddard and J. E. Greaves, *The Composition of Summer Range Plants in Utah* (Bull. 305, Utah Agric. Exp. Stn., Logan, 1942).

5. C. M. Aldous, "A winter study of mule deer in Nevada," *J. Wildl. Manage.* 9 (1945): 145–154. We cited this publication earlier in a historic context concerning the population growth of mule deer during the first half of the twentieth century, but it is also one of the earliest articles dealing with the nutrition of mule deer.

6. Perhaps the most noted and somewhat controversial example of this is J. G. Nagy, H. W. Steinhoff, and G. M. Ward, "Effect of essential oils of sagebrush on deer rumen microbial function," *J. Wildl. Manage.* 28 (1964): 785–790.

7. Arthur D. Smith, "Sagebrush as a winter feed for deer," *J. Wildl. Manage.* 14 (1950): 285–289.

8. Arthur D. Smith, "Digestibility of some native forages for mule deer," *J. Wildl. Manage.* 16 (1952): 309–312.

9. Arthur D. Smith, "Nutritive value of some browse plants in winter," *J. Range Manage.* 10 (1957): 162–164.

10. Arthur D. Smith, "Consumption of native forage species by captive mule deer during summer," *J. Range Manage.* 6 (1953): 30–37.

11. See, for example, Justin G. Smith, "Food habits of mule deer in Utah," *J. Wildl. Manage.* 16 (1952): 148–155.

12. Harold D. Bissell, Bruce Harris, Helen Strong, and Frank James, "The digestibility of certain natural and artificial foods eaten by deer in California," *Calif. Fish and Game* 41 (1955): 57–78; Herbert L. Hagen, "Nutritive value for deer of some forage plants in the Sierra Nevada," *Calif. Fish and Game* 39 (1953): 163–175. Hagen collected antelope bitterbrush from a heavily browsed stand near Lake Almanor in northern Cali-

fornia. Apparently, a crude protein determination (13.7 percent) was the only test he performed on this browse.

13. Arthur D. Smith, "Feeding deer on browse species during the winter," *J. Range Manage.* 10 (1950): 130–132.

14. R. W. Lassen, C. M. Ferrel, and H. Leach, "Food habits, productivity, and condition of the Doyle mule deer herd," *Calif. Fish and Game* 38 (1952): 211–224.

15. H. D. Bissell and H. Strong, "The crude protein variations in the browse diet of California deer," *Calif. Fish and Game* 41 (1955): 145–155.

16. Arthur D. Smith and Richard L. Hubbard, "Preference ratings for winter deer forages from northern Utah ranges based on browsing time and forage consumption," *J. Range Manage.* 7 (1954): 262–265. This is the Richard Hubbard who conducted extensive restoration trials with antelope bitterbrush in northeastern California as a scientist with the USDA Forest Service in Berkeley, California.

17. See, for example, Bruce L. Welch, Stephen B. Monsen, and Nancy L. Shaw, "Nutritive value of antelope and desert bitterbrush," pp. 173–175 in *Proceedings of the Symposium on Research and Management of Bitterbrush and Cliffrose in Western North America,* ed. A. R. Tiedemann and K. L. Johnson (Gen. Tech. Rep. 152, USDA, Forest Serv., Ogden, Utah, 1983).

18. See, for example, *Nutrient Requirements of Domestic Animals,* no. 5: *Nutrient Requirements of Sheep* (Publ. 1193, Natl. Res. Council; Washington, D.C.: National Academy of Sciences, 1964).

19. Welch et al., "Nutritive value of antelope and desert bitterbrush."

20. H. R. Maynard, J. K. Loosli, H. F. Hintz, and R. G. Warner, *Animal Nutrition,* 7th ed. (New York: McGraw-Hill, 1979).

21. H. R. Leach, "Food habits of the Great Basin deer herds of California," *Calif. Fish and Game* 42 (1956): 143–308; Paul T. Tueller, "Food habits and nutrition of mule deer on Nevada ranges" (Nev. Agric. Exp. Stn., Reno, 1979); D. E. Medin, unpublished data on file at the USDA, Forest Serv., Shrub Sci. Lab., Provo, Utah (as cited by Welch et al.).

22. B. L. Welch and D. Andrus, "Rose hips—a possible energy food for wintering mule deer" (Res. Note 221, USDA, Forest Serv., Ogden, Utah, 1977).

23. For an excellent discussion of the key species concept, see Arthur D. Smith, "Determining common use grazing capacities by application of the key species concept," *J. Range Manage.* 18 (1965): 196–201.

24. Galen C. Burrell, "Winter diets of mule deer in relation to bitterbrush abundance," *J. Range Manage.* 35 (1982): 508–510.

25. For a discussion of the use of mule deer fecal samples, see Henry L. Short and Elmer E. Remmenga, "Use of fecal cellulose to estimate plant tissue eaten by deer," *J. Range Manage.* 18 (1965): 139–144.

26. V. B. Richens, "Characteristics of mule deer herds and their range in northeastern Utah," *J. Wildl. Manage.* 31 (1967): 651–666; B. T. Wilkens, "Range use, food habits, and agricultural relationships of mule deer, Bridger Mountains, Montana," *J. Wildl. Manage.* 21 (1957): 159–169.

27. R. H. Hansen and B. L. Dearden, "Winter foods of mule deer in Piceance Basin, Colorado," *J. Range Manage.* 28 (1975): 298–300.

28. Welch et al., "Nutritive value of antelope and desert bitterbrush," provides the citations, most of which we have already cited (Arthur Smith, Bissell et al., and Dietz); an addition is P. J. Urness, A. D. Smith, and R. K. Watkins, "Comparison of *in vivo* and *in vitro* dry matter digestibility of deer food species," *J. Range Manage.* 28 (1977): 419–421.

29. C. W. Cook, "Comparative nutritive values of forbs, grasses, and shrubs," pp. 303–310 in *Wildland Shrubs—Their Biology and Utilization,* ed. C. M. Blaisdell and J. P. Gooden (Gen. Tech. Rep. 1, USDA, Forest Serv., Ogden, Utah, 1972).

30. Welch et al., "Nutritive value of antelope and desert bitterbrush."

31. O. Eugene Hickman, "Seasonal trends in the nutritive content of important range forage species near Silver Lake, Oregon" (Res. Pap. 187, USDA, Forest Serv., Portland, Ore., 1975).

32. Carl L. Wambolt, W. Wyatt Fraas, and M. R. Frisina, "Variation in bitterbrush (*Purshia tridentata*) crude protein in southwestern Montana," *Great Basin Nat.* 56 (1996): 205–210.

33. Welch et al., "Nutritive value of antelope and desert bitterbrush."

34. J. M. Alderfer, "A taxonomic study of bitterbrush (*Purshia tridentata*) in Oregon" (master's thesis, Ore. State Univ., Corvallis, 1976).

35. J. A. Young and R. A. Evans, "Reciprocal common garden studies of the germination of seeds of big sagebrush (*Artemisia tridentata*)," *Weed Sci.* 37 (1989): 319–325.

36. Interstate Deer Herd Committee, "Second progress report on the cooperative study of the interstate deer herd and its range," *Calif. Fish and Game* 33 (1947): 287–314.

37. Welch et al., "Nutritive value of antelope and desert bitterbrush."

38. Hickman, "Seasonal trends in nutritive content."

39. J. E. Dealy, "Bitterbrush nutrition levels under natural and thinned ponderosa pine" (Res. Note 33, USDA, Forest Serv., Portland, Ore., 1966).

40. Cook, "Nutritive value of antelope and desert bitterbrush."

41. Roland C. Kufeld, O. C. Wallmo, and Charles Feddema, "Foods of the Rocky Mountain mule deer" (Res. Pap. 111, USDA, Forest Serv., Fort Collins, Colo., 1973).

42. G. Bryan Harry, "Winter food habits of moose in Jackson Hole, Wyoming," *J. Wildl. Manage.* 21 (1957): 53–57.

43. Dale R. McCullough and Edward R. Schneegas, "Winter observations on the Sierra Nevada bighorn sheep," *Calif. Fish and Game* 52 (1966): 68–84; Carol M. Ferrel and Howard R. Leach, "Food habits of the pronghorn antelope in California," *Calif. Fish and Game* 36 (1950): 21–26.

44. S. McKeever and R. L. Hubbard, "Use of desert shrubs by jackrabbits in northeastern California," *Calif. Fish and Game* 46 (1960): 271–277.

45. Dennis D. Austin and Philip J. Urness, "Consumption of fresh alfalfa hay by mule deer and elk," *Great Basin Nat.* 47 (1987): 100–102.

Chapter Nine. Insects and Plant Diseases

1. M. M. Furniss, "A preliminary list of insects and mites that infest some important browse plants of western big game" (Res. Note 155, USDA, Forest Serv., Ogden, Utah, 1972).

2. M. M. Furniss, "Entomology of antelope bitterbrush," pp. 164–172 in *Proceedings of the Symposium on Research and Management of Bitterbrush and Cliffrose in Western North America*, ed. A. R. Tiedemann and K. L. Johnson (Gen. Tech. Rep. 152, USDA, Forest Serv., Ogden, Utah, 1983).

3. R. B. Ferguson, M. M. Furniss, and J. V. Basile, "Insects destructive to bitterbrush flowers and seeds in southwestern Idaho," *J. Econ. Entomol.* 47 (1963): 268–272; J. V. Basile, R. B. Ferguson, and M. M. Furniss, "Six-legged seed eaters," *Idaho Wildl. Rev.* 17, no. 3 (1964): 5–7.

4. Basile et al., "Six-legged seed eaters."

5. Furniss, "Entomology of antelope bitterbrush"; Basile et al., "Six-legged seed eaters"; and Ferguson et al., "Insects destructive to bitterbrush." Obviously, insects can be very difficult to identify.

6. Furniss, "Entomology of antelope bitterbrush."

7. J. V. Basile and R. B. Ferguson, "Say stink bug destroys bitterbrush seed," *J. Range Manage.* 17 (1964): 153–154.

8. E. J. Woolfolk, "Semi-annual report to Western Browse Revegetation Committee," *Western Browse Res.* 5.1 (1959): n.p.; personal communication from R. L. Hubbard (1961) as cited by Ferguson et al., "Insects destructive to bitterbrush."

9. Furniss, "Preliminary list."

10. Ibid.

11. Edwin C. Clark, "The Great Basin tent caterpillar in relation to bitterbrush in California," *Calif. Fish and Game* 42 (1956): 131–142.

12. Jeff Knight, state entomologist with the Nevada Division of Agriculture, suggests the valid scientific names are *Malacosoma californicum californicum* and *M. californicum fragile*.

13. F. P. Keen, *Insect Enemies of Forest* (Misc. Publ. 273, USDA, Washington, D.C., 1952).

14. Russel G. Mitchell, "Seasonal history of the western tent caterpillar (Lepidoptera: Lasiocampidae) on bitterbrush and currant in central Oregon," *J. Econ. Entomol.* 83 (1990): 1492–1494.

15. Clark, "Great Basin tent caterpillar."

16. Ibid.

17. Ibid.

18. Ibid.

19. Mitchell, "Seasonal history of western tent caterpillar."

20. Clark, "Great Basin tent caterpillar."

21. Personal communication from Bill Phillips, retired range conservationist, Bureau of Land Management, Susanville, Calif., November 22, 1992.

22. Mitchell, "Seasonal history of western tent caterpillar."

23. Clark, "Great Basin tent caterpillar."

24. Edwin C. Clark and Clarence G. Thompson, "The possible use of microorganisms in the control of the Great Basin tent caterpillar," *J. Econ. Entomol.* 47 (1954): 268–272.

25. We are not certain that *Malacosoma californicum* and *M. fragilis* are not synonyms or that the correct taxon should be *M. californicum* subsp. *fragilis* for the Great Basin tent caterpillar. Apparently *Malacosoma* includes several highly variable species.

26. Ferguson et al., "Insects destructive to bitterbrush"; Furniss, "Entomology of antelope bitterbrush."

27. Malcolm M. Furniss and Gordin A. Van Epps, "Bionomics and control of the walnut spanworm, *Phigalia plumogeraria* (Hulst), on bitterbrush in Utah," *Great Basin Nat.* 41 (1981): 290–297.

28. Furniss, "Entomology of antelope bitterbrush."

29. M. M. Furniss and W. F. Barr, *Bionomics of* Anacamptodes clivinaria profanata *(Lepidoptera: Geometridae) on Mountain Mahogany in Idaho* (Res. Bull. 73, Agric. Exp. Stn., Univ. Idaho, Moscow, 1967).

30. M. M. Furniss and J. A. E. Knopf, *Western Tussock Moth* (Pest Leaflet 120, USDA, Forest Serv., Washington, D.C., 1971).

31. Furniss, "Entomology of antelope bitterbrush."

32. Ibid.

33. Malcolm M. Furniss and William F. Barr, *Insects Affecting Important Native Shrubs of the Northeastern United States* (Gen. Tech. Rep. INT-19, USDA, Forest Serv., Intermountain Forest and Range Exp. Stn., Ogden, Utah, 1975).

34. Ralph C. Holmgren, "Progress report for 1953: Cooperative research program for revegetating the deer winter ranges on the Payette and Boise River drainages" (USDA, Forest Serv., Ogden, Utah, 1954).

35. R. L. Hubbard, "Bitterbrush seedlings destroyed by cutworms and wireworms" (Res. Note 114, USDA, Forest Serv., Berkeley, Calif., 1956).

36. Furniss, "Entomology of antelope bitterbrush."

37. Furniss and Krebill, "Insects and diseases of shrubs."

38. "Forest pest conditions in California" (Calif. Forest Pest Action Council, Calif. Div. Forest., Sacramento, 1970).

39. David L. Nelson, "Susceptibility of antelope bitterbrush to seedbed damping-off disease," pp. 117–121 in *Proceedings of a Symposium on Seed and Seedbed Ecology of Rangeland Plants,* ed. Gary W. Frasier and Raymond A. Evans (USDA, ARS, Washington, D.C., 1987).

40. R. A. Peterson, "Comparative effects of seed treatments upon seedling emergence of seven browse species," *Ecology* 34 (1953): 778–785.

41. R. C. Holmgren, "A comparison of browse species for revegetation of big game winter ranges in southeastern Idaho" (Res. Pap. 33, USDA, Forest Serv., Ogden, Utah, 1954); R. B. Brown and C. F. Martinsen, *Browse Planting for Big Game in the State of Washington* (Biol. Bull. 690, Wash. State Game Dep., Olympia, 1959).

42. Nelson, "Susceptibility of antelope bitterbrush."

43. D. L. Nelson, "Seedborne fungal pathogens of bitterbrush (*Purshia tridentata*)," *Phytopathol. Abstr.* 75 (1985): 1368.

44. Ibid.

45. Alfred C. Tegtoff, "Known distribution of *Fomes annosus* in the Intermountain region," *Plant Dis. Rep.* 57 (1973): 407–410.

Chapter Ten. Wildfire Relations

1. S. W. Barrett, "Indians and fire," *Western Wildlands* 6.3 (1980): 17–21; G. E. Gruel, "Fire on the early western landscape," *Northwest Sci.* 59 (1985): 97–107.

2. J. A. Young and R. A. Evans, "Population dynamics after wildfires in sagebrush grasslands," *J. Range Manage.* 31 (1978): 283–289.

3. J. A. Young, R. A. Evans, and D. E. Palmquist, "Soil surface characteristics and emergence of big sagebrush seedlings," *J. Range Manage.* 43 (1990): 358–367; J. A. Young and R. A. Evans, "Germinability of seed reserves in a big sagebrush community," *Weed Sci.* 23 (1975): 358–364; J. A. Young and R. A. Evans, "Reciprocal common garden studies of the germination of seeds of big sagebrush (*Artemisia tridentata*)," *Weed Sci.* 37 (1989): 319–325; J. A. Young and R. A. Evans, "Dispersal and germination of big sagebrush (*Artemisia tridentata*) seeds," *Weed Sci.* 37 (1989): 201–206.

4. J. A. Young and R. A. Evans, "Population dynamics of green rabbitbrush in disturbed sagebrush communities," *J. Range Manage.* 27 (1974): 127–132.

5. August L. Hormay, "Bitterbrush in California" (Res. Note 34, USDA, Forest Serv., Berkeley, Calif., 1943).

6. Eamor C. Nord, "Autecology of bitterbrush in California," *Ecol. Monogr.* 35 (1965): 307–334; J. P. Blaisdell, *Ecological Effects of Planned Burning of Sagebrush-Grass Ranges on the Upper Snake River Plains* (Tech. Bull. 1075 USDA, Forest Serv., Washington, D.C., 1953); C. M. Countryman and D. R. Cornelius, "Some effects of fire on a perennial range type," *J. Range Manage.* 10 (1957): 39–41; J. F. Pechanec, G. Steward, and J. P. Blaisdell, *Sagebrush Burning — Good and Bad* (Farmers Bull. 1948, USDA, Forest Serv., Washington, D.C., 1954); Eamor C. Nord, "Bitterbrush ecology—some recent findings" (Res. Note 148, USDA, Forest Serv., Berkeley, Calif., 1959).

7. J. P. Blaisdell and W. F. Mueggler, "Sprouting of bitterbrush (*Purshia tridentata*) following burning or top removal," *Ecology* 37 (1956): 365–369.

8. Harold Weaver, "Fire as an ecological and silvicultural factor in the ponderosa pine region of the Pacific slope," *J. Forest.* 41 (1943): 7–18.

9. See R. S. Rummell, "Some effects of livestock grazing on ponderosa pine forest and range in central Washington," *Ecology* 32 (1951): 594–607.

10. S. F. Arno, "Historical role of fire on the Bitterroot National Forest" (Res. Pap. 187, USDA, Forest Serv., Ogden, Utah, 1976).

11. C. H. Driver, V. A. G. Wiston, and H. F. Gobble, "The fire ecology of bitterbrush—a proposed hypothesis," pp. 204–208 in *Proceedings of the 6th Conference of Forest Meteorology*, ed. R. R. Martin et al. (Washington, D.C.: Society of American Foresters, 1980).

12. A. H. Johnson and G. A. Smothers, "Fire history and ecology, Lava Beds National Monument," pp. 103–115 in *Tall Timbers Fire Ecology Contributions,* vol. 2 (Tall Timbers Res. Sta., Tallahassee, Fla., 1976).

13. S. F. Arno and K. M. Sneck, *A Method for Determining Fire History in Coniferous Forest of the Mountain West* (Gen. Tech. Rep. 42, USDA, Forest Serv., Ogden, Utah, 1977); Henry A. Wright, *The Effect of Fire on Vegetation in Ponderosa Pine Forest* (Publ. 9-199, College Agric. Sci., Tex. Tech. Univ., Lubbock, 1978).

14. T. N. Johnson Jr., "One-seeded juniper invasion of northern Arizona," *Ecol. Monogr.* 32 (1962): 187–207.

15. J. A. Young and J. D. Budy, "Historical use of Nevada's pinyon-juniper," *J. Forest. Hist.* 23 (1979): 112–121.

16. See, for example, J. W. Burkhardt and E. W. Tisdale, "Nature and successional status of western juniper vegetation in Idaho," *J. Range Manage.* 22 (1969): 264–270.

17. A. D. Brunner and D. A. Klebenow, "Predicting success of prescribed fires in pinyon-juniper woodlands in Nevada" (Res. Pap. 84, USDA, Forest Serv., Ogden, Utah, 1979).

18. See, for example, P. J. Mehringer, S. R. Arno, and K. L. Peterson, "Postglacial history of Lost Trail Pass Bog, Bitterroot Mountains, Montana," *Arctic Alpine Res.* 9 (1977): 345–368.

19. J. A. Young, R. A. Evans, and C. Rimby, "Weed control and revegetation following western juniper (*Juniperus occidentalis*) control," *Weed Sci.* 33 (1985): 513–517.

20. R. F. Daubenmire described lignotubers in his book *Plants and Environment* (New York: Wiley and Sons, 1974).

21. Carol L. Rice, "A literature review of the fire relationships of antelope bitterbrush," pp. 256–264 in *Proceedings of the Symposium on Research and Management of Bitterbrush and Cliffrose in Western North America,* ed. A. R. Tiedemann and K. L. Johnson (Gen. Tech. Rep. 152, USDA, Forest Serv., Ogden, Utah, 1983).

22. Johnson and Smothers, "Fire history and ecology."

23. R. Martin, "Fire manipulation and effect in western juniper (*Juniperus occidentalis* Hook.)," pp. 121–136 in *Proceedings of the Western Juniper Ecology and Management Workshop* (Gen. Tech Rep. 74, USDA, Forest Serv., Portland, Ore., 1978).

24. R. F. Blank, Fay Allen, and J. A. Young, "Growth and elemental content of several sagebrush-steppe species in unburned and post-wildfire soil and plant effects on soil attributes," *Plant Soil* 164 (1994): 35–41. This is one example of the numerous papers published by Robert Blank concerning the influence of burning on the chemistry of seedbeds.

25. R. Martin and J. Dell, *Planning for Prescription Burning in the Inland Northwest* (Gen. Tech. Rep. 76, USDA, Forest Serv., Portland, Ore., 1978).

26. Driscoll, "Sprouting bitterbrush in central Oregon."

27. Personal communication cited by Robert Martin and Charles H. Driver in "Factors affecting antelope bitterbrush reestablishment following fire," pp. 266–279 in Tiedemann and Johnson 1983.

28. L. R. Green, *Burning by Prescription in Chaparral* (Gen. Tech. Rep. 51, USDA, Forest Serv., Berkeley, Calif., 1981).

29. Robert Martin, David Frewing, and James McClanahan, "Average biomass of four northwest shrubs by fuel size class and crown cover" (Res. Note 374, USDA, Forest Serv., Portland, Ore., 1981).

30. E. C. Nord and C. M. Countryman, "Fire relationships," in *Shrubs—Their Biology and Utilization,* ed. C. M. McKell, J. P. Blaisdell, and J. R. Goodin (Gen. Tech. Rep. 1, USDA, Forest Serv., Ogden, Utah, 1972).

31. C. W. Philpot, "Seasonal changes in heat content and ether extractive content of chamise" (Res. Pap. 61, USDA, Forest Serv., Ogden, Utah, 1969).

32. Glenn Adams, "Results of range/wildlife prescribed burning on the Fort Rock Ranger District in central Oregon" (Fuels Manage. Notes 6, USDA, Forest Serv., Pac. Northwest Reg., Portland, Ore., 1980).

33. W. D. Billings, "The environmental complex in relation to plant growth and distribution," *Q. Rev. Biol.* 27 (1952): 251–265.

34. Blaisdell and Mueggler, "Sprouting of bitterbrush"; Richard S. Driscoll, "Sprouting bitterbrush in central Oregon," *Ecology* 44 (1963): 820–821.

35. A. Starker Leopold, "Deer in relation to plant succession," *J. Forest.* 48 (1950): 675–678.

36. Blaisdell, *Ecological Effects of Planned Burning.*

37. Blaisdell and Mueggler, "Sprouting of bitterbrush."

38. Fred J. Wagstaff, "Impact of the 1975 Wallsburg fire on antelope bitterbrush (*Purshia tridentata*)," *Great Basin Nat.* 40 (1980): 299–302.

39. John G. Cook, Terry J. Hersey, and Larry L. Irwin, "Vegetative response to burning on Wyoming mountain-shrub big game ranges," *J. Range Manage.* 47 (1994): 296–302.

40. W. W. Fraas, C. L. Wambolt, and M. R. Frisina, "Prescribed fire effects on a bitterbrush–mountain big sagebrush–bluebunch wheatgrass community," pp. 212–216 in *Proceedings of a Symposium on Ecology and Management of Riparian Shrub Communities,* ed. W. P. Clary, E. D. McArthur, D. Bedunah, and C. L. Wambolt (Gen. Tech. Rep. 289, USDA, Forest Serv., Ogden, Utah, 1992).

41. Martin and Driver, "Factors affecting antelope bitterbrush reestablishment."

42. Robert Fay Wagle, "Early growth in bitterbrush and its relation to environment" (Ph.D. diss., Univ. Calif., Berkeley, 1958).

43. Driver et al., "Fire ecology of bitterbrush."

44. B. R. McConnell and J. G. Smith, "Influence of grazing on age-yield interactions in bitterbrush," *J. Range Manage.* 30 (1977): 91–93; Robert G. Clark, "Seasonal response of bitterbrush to burning and clipping in eastern Oregon" (M.S. thesis, Ore. State Univ., Corvallis, 1979).

45. Martin and Driver cited the following papers: Blaisdell, *Ecological Effects of Planned Burning;* Blaisdell and Mueggler, "Sprouting of bitterbrush"; Nord, "Autecology of bitterbrush in California"; J. F. Pechanec, A. P. Plummer, J. H. Robertson, and A. C. Hull Jr., *Sagebrush Control on Rangelands* (Agric. Handb. 227, USDA, Washington, D.C., 1965); H. A. Wright and C. M. Britton, "Fire effects on vegetation in western

rangeland communities," pp. 35–41 in *Use of Prescribed Burning in Western Woodlands and Range Ecosystems* (Utah Agric. Exp. Stn., Logan, 1976); R. E. Martin and J. D. Dell, *Planning for Prescribed Burning in the Inland Northwest* (Gen. Tech. Rep. 76, USDA, Forest Serv., Portland, Ore., 1978); R. E. Martin and A. H. Johnson, "Fire management of Lava Beds National Monument," in *Proceedings of the First Conference on Scientific Research in the National Parks,* vol. 2 (USDI, Natl. Park Serv., Washington, D.C., 1979); Adams, "Results of range/wildlife prescribed burning"; C. M. Olson, R. E. Martin, and A. H. Johnson, "Fire effects on vegetation in sagebrush-grass and ponderosa pine communities," pp. 373–388 in *Proceedings of the Second Conference on Scientific Research in the National Parks,* vol. 10 (1981); L. A. Volland and J. D. Dell, "Fire effects on Pacific Northwest forest and range vegetation" (R-6, USDA, Forest Serv., Portland, Ore., 1981); and Murray, "Response of antelope bitterbrush to burning and spraying in southeastern Idaho," pp. 142–152 in Tiedemann and Johnson 1983. Except for the Blaisdell papers, most of these are review articles that repeat previous assumptions without original data from the field.

46. R. E. Martin, "Antelope bitterbrush seedling establishment following prescribed burning in the pumice zone of the southern Cascade Mountains," pp. 82–90 in Tiedemann and Johnson 1983.

47. Ibid.

48. Blaisdell, *Ecological Effects of Planned Burning.*

49. Murray, "Response of antelope bitterbrush to burning and spraying in southeastern Idaho."

50. G. D. Pickford, "The influence of continued heavy grazing and of promiscuous burning on spring-fall ranges in Utah," *Ecology* 13 (1932): 169–171.

51. Pechanec et al., *Sagebrush Control on Rangelands.*

52. Eric R. Loft and John W. Menke, "Evaluation of fire effects on mule deer habitat in Lassen County" (final report, Calif. Dep. Fish and Game, Hill Bill Contract FG1C-2090, Sacramento, 1990); Douglas Updike, E. R. Loft, and F. A. Hall, "Wildfires on big sagebrush/antelope bitterbrush range in northeastern California: Implications for deer populations," pp. 41–46 in *Proceedings of a Symposium on Shrub Die-off and Other Aspects of Shrub Biology and Management* (Gen. Tech. Rep. 276, USDA, Forest Serv., Ogden, Utah, 1991).

53. Driscoll, "Sprouting bitterbrush in central Oregon."

54. Robert Clark, Carlton Britton, and Forrest Sneva, "Mortality of bitterbrush after burning and clipping in eastern Oregon," *J. Range Manage.* 35 (1982): 711–714; Clark, "Seasonal response of bitterbrush."

55. The authors cited the classic papers on bitterbrush-rodent relations: N. E. West, "Rodent-influenced establishment of ponderosa pine and bitterbrush seedlings in central Oregon," *Ecology* 49 (1968): 1009–1011; and R. J. Sherman and W. W. Chilcote, "Spatial and chronological patterns of *Purshia tridentata* as influenced by *Pinus ponderosa,*" *Ecology* 53 (1972): 294–298.

56. Charles H. Driver, "Potentials for the management of bitterbrush habitats by the use of prescribed fire," pp. 137–141 in Tiedemann and Johnson 1983. Also see Driver et al., "Fire ecology of bitterbrush."

Chapter Eleven. The Role of Nitrogen

1. J. O. Klemmedson and R. B. Ferguson, "Response of bitterbrush seedlings to nitrogen and moisture on a granitic soil," *Soil Sci. Soc. Am. Proc.* 33 (1969): 962–966.

2. See, for example, H. Jenny, J. Vlamis, and W. E. Martin, "Greenhouse assay of fertility of California soils," *Hilgardia* 20 (1950): 1–8.

3. Klemmedson and Ferguson, "Response of bitterbrush seedlings."

4. J. O. Klemmedson and R. B. Ferguson, "Effect of sulfur deficiency on yield and nitrogen content in bitterbrush (*Purshia tridentata*) seedlings on granitic soils," *Soil Sci. Soc. Am. Proc.* 37 (1973): 947–951.

5. R. F. Wagle and J. Vlamis, "Nutrient deficiencies in two bitterbrush soils," *Ecology* 42 (1961): 745–752; based on R. F. Wagle, "Early growth in bitterbrush and its relation to environment" (Ph.D. diss., Univ. Calif., Berkeley, 1958).

6. Jenny developed this replacement series for pot-testing soils based on the procedures first proposed by E. A. Mitscherlich in *Die Bestimmung des Dungebedurfnsses des Bodens* (Berlin: P. Parey, 1930).

7. See, for example, G. Bond, "An isotopic study of the fixation of nitrogen associated with nodulated plants of *Alnus, Myrica,* and *Hippophae,*" *J. Exp. Bot.* 6 (1955): 303–311; J. C. Gardner and G. Bond, "Observations on the root nodules of *Shepherdia,*" *Can. J. Bot.* 35 (1957): 305–314; R. L. Crocker and J. Major, "Soil development in relation to vegetation and surface age at Glacier Bay, Alaska," *J. Ecol.* 43 (1955): 427–428; C. R. Quick, "Effects of snowbrush on the growth of Sierra gooseberry," *J. Forest.* 42 (1944): 827–832.

8. See, for example, W. B. Bottomley, "The root nodules of *Ceanothus americanus,*" *Ann. Bot.* (London) 29 (1915): 605–610.

9. B. A. Dickson and R. L. Crocker, "A chronosequence of soils and vegetation near Mount Shasta, California," *J. Soil Sci.* 4 (1953): 123–141. This is the same Crocker who two years later published with Jack Major of the University of California at Davis the paper about nodulated shrubs and plant succession at Glacier Bay, Alaska.

10. Wagle and Vlamis, "Nutrient deficiencies."

11. B. Leman and C. C. Pittack, "Pilot study: Pruning and fertilization of bitterbrush on the Swakane Game Range" (unpublished report on file with Cheland Co. PUD, Wenatchee, Wash., 1969, as cited in A. R. Tiedemann, "Response of bitterbrush and associated plant species to broadcast nitrogen, phosphorous, and sulfur fertilization," pp. 240–253 in *Proceedings of the Symposium on Research and Management of Bitterbrush and Cliffrose in Western North America,* ed. A. R. Tiedemann and K. L. Johnson [Gen. Tech. Rep. 152, USDA, Forest Serv., Odgen, Utah, 1983]).

12. M. A. Bayoumi and A. D. Smith, "Response of big game winter range vegetation to fertilization," *J. Range Manage.* 29 (1976): 44–48.

13. Tiedemann, "Response of bitterbrush and associated plant species."

14. J. O. Klemmedson, "Ecological importance of actinomycete-nodulated plants in the western United States," *Bot. Gaz.* 140 (1979): S91–S96. Klemmedson used a list published in J. H. Becking, "Dinitrogen-fixing associations in higher plants other than

legumes," pp. 185–275 in *A Treatise on Dinitrogen Fixation,* ed. R. W. F. Hardy and W. S. Silver, sec. 3: Biology (New York: Wiley and Sons, 1977).

15. E. D. McArthur, B. C. Giunta, and A. P. Plummer, "Shrubs for restoration of depleted ranges and disturbed areas," *Utah Sci.* 35 (1974): 28–33.

16. Klemmedson, "Ecological importance of actinomycete-nodulated plants."

17. R. B. Farnsworth, E. M. Rommey, and A. Wallace, "Implications of symbiotic nitrogen fixation by desert plants," *Great Basin Nat.* 36 (1976): 65–80. A second often cited review is E. K. Allen and O. N. Allen, "Biological aspects of symbiotic nitrogen fixation," pp. 43–105 in *Encyclopedia of Plant Physiology,* vol. 8, ed. W. Ruhland (Berlin: Springer-Verlag, 1958).

18. W. G. McGinnies, "Continental aspects of shrub distribution, utilization, and potentials: North America," pp. 55–66 in *Wildland Shrubs — Their Biology and Utilization,* ed. C. M. McKell, J. P. Blaisdell, and J. R. Goodin (Gen. Tech. Rep. 1, USDA, Forest Serv., Odgen, Utah, 1972).

19. E. C. Nord, "Autecology of bitterbrush in California," *Ecol. Monogr.* 36 (1965): 307–334.

20. See, for example, Edmundo Garcia-Moya and Cyrus M. McKell, "Contribution of shrubs to the nitrogen economy of a desert-wash plant community," *Ecology* 51 (1970): 81–88.

21. See, for example, D. B. Lawrence, R. E. Schoenike, A. Quispel, and G. Bond, "The role of *Dryas drummondii* in vegetation development following ice recession at Glacier Bay, with special reference to its nitrogen fixation roots," *J. Ecol.* 55 (1967): 783–813.

22. The classic paper often cited to support this statement is Crocker and Major, "Soil development at Glacier Bay."

23. S. R. Webster, C. T. Youngberg, and A. G. Wollum II, "Fixation of nitrogen by bitterbrush (*Purshia tridentata* [Pursh] D.C.)," *Nature* 216 (1967): 392–393.

24. See, for example, J. H. Becking, "Plant-endophyte symbiosis in non-leguminous plants," *Plant Soils* 32 (1970): 611–654.

25. J. H. Becking, "Actinomycete symbioses in non-legumes," pp. 581–591 in *Proceedings of the First International Symposium on Nitrogen Fixation,* ed. W. E. Newton and C. J. Nyman (Pullman: Wash. State Univ. Press, 1976).

26. J. H. Becking, "Frankiaceae fam. nov. (Actinomycetales) with one new combination and six species of the genus *Frankia* Brunchorst 1886, 174," *Int. J. Syst. Bacteriol.* 20 (1970): 201–202.

27. R. E. Hoppel and A. G. Wollum II, "Histological studies of ectomycorrhizae and root nodules from *Cercocarpus montanus* and *Cercocarpus pauidentalis,*" *Can. J. Bot.* 49 (1971): 1315–1318.

28. R. J. Krebill and J. M. Muir, "Morphological characterization of *Frankia purshiae,* the endophyte in root nodules of bitterbrush," *Northwest Sci.* 48 (1974): 266–268.

29. G. Bond, "Observations on the root nodules of *Purshia tridentata,*" *Proc. Roy. Soc. Lond.* B 193 (1976): 127–135.

30. David A. Dalton and Donald B. Zobel, "Ecological aspects of nitrogen fixation by *Purshia tridentata,*" *Plant Soils* 48 (1977): 57–80.

31. C. T. Youngberg and A. G. Wollum II, "Non-leguminous symbiotic nitrogen fixation," pp. 383–395 in *Tree Growth and Forest Soils,* ed. C. T. Youngberg (Corvallis: Ore. State Univ. Press, 1970).

32. A. G. Wollum II, C. T. Youngberg, and F. W. Chichester, "Relation of previous timber stand age to nodulation of *Ceanothus velutinus,*" *Forest Sci.* 14 (1968): 114–118.

33. I. C. Gardner and G. Bond, "Observations on the root nodules of *Shepherdia,*" *Can. J. Bot.* 35 (1957): 305–314.

34. G. Bond, "Fixation of nitrogen by higher plants other than legumes," *Ann. Rev. Plant Physiol.* 18 (1967): 107–126.

35. T. L. Righetti, "Soil factors limiting nodulation and nitrogen fixation in bitterbrush (*Purshia*)" (Ph.D. diss., Univ. Calif., Davis, 1980).

36. Timothy L. Righetti and Donald N. Munns, "Nodulation and nitrogen fixation in *Purshia*: Inoculation responses and species comparison," *Plant Soils* 65 (1982): 383–396.

37. Righetti, "Soil factors limiting nodulation and nitrogen fixation."

38. Timothy L. Righetti and D. N. Munns, "Soil factors limiting nodulation and nitrogen fixation in *Purshia,*" pp. 395–407 in *Genetic Engineering of Symbiotic Nitrogen Fixation and Conservation of Fixed Nitrogen,* ed. J. M. Lyons, R. C. Valentine, D. A. Phillips, D. W. Rams, and R. C. Huffaker (New York: Plenum, 1981).

39. Timothy L. Righetti, Carolyn H. Chard, and D. N. Munns, "Opportunities and approaches for enhancing nitrogen fixation in *Purshia, Cowania,* and *Fallugia,*" pp. 214–223 in Tiedemann and Johnson 1983.

40. D. Callaham, P. D. Tredici, and J. G. Torrey, "Isolation and cultivation *in vitro* of the actinomycete causing root nodulation in *Comptonia,*" *Science* 199 (1978): 899–902.

41. Righetti et al., "Opportunities and approaches for enhancing nitrogen fixation."

42. Timothy L. Righetti and Donald N. Munns, "Nodulation and nitrogen fixation in cliffrose (*Cowania mexicana* var. *stansburiana* [Torr.] Jeps.)," *Plant Physiol.* 65 (1980): 411–412.

43. Righetti et al., "Opportunities and approaches for enhancing nitrogen fixation."

44. David L. Nelson and Patti L. Schuttler, "Histology of *Cowania stansburiana* actiorhizae," *Northwest Sci.* 58 (1984): 49–56.

45. C. B. Perry, J. C. Stutz, and T. L. Righetti, "*Cowania subintegra* (Rosaceae): A new actinorhizal species," p. 701 in *Nitrogen Fixation Research Progress,* ed. H. J. Evans, P. J. Bottomley, and W. E. Newton (Dordrecht, Netherlands: Nijhoff, 1985).

46. David L. Nelson, "Occurrence and nature of actinorhizae in *Cowania stansburiana* and other Rosaceae," pp. 225–239 in Tiedemann and Johnson 1983.

47. Ibid.

48. Wolfgang Honerlage, Dittmar Hahn, Kornelia Zepp, Joseph Zeyer, and Philippe Norman, "A hypervariable 23S rRNA region provides a discriminating target for specific characterization on uncultured and cultured *Frankia,*" *Syst. Appl. Microbiol.* 17 (1994): 433–443.

49. Stephen E. Williams, "Vesicular-arbuscular mycorrhizae associated with actinomycete-nodulated shrubs, *Cercocarpus montanus* Raf. and *Purshia tridentata* (Pursh) DC.," *Bot. Gaz.* 140 (Suppl.) (1979): S115–S119.

50. J. A. Young, R. R. Blank, and W. S. Longland, "Nitrogen enrichment-immobilization to control succession in arid land plant communities," *J. Arid Land Stud.* 58 (1996): 57–60.

51. J. A. Young, Charlie D. Clements, and Robert R. Blank, "Influence of nitrogen on antelope bitterbrush seedling establishment," *J. Range Manage.* 50 (1997): 536–540.

Chapter Twelve. Purshia *Management*

1. G. D. Pickford, "The influence of the continued heavy grazing and of promiscuous burning on spring-fall ranges in Utah," *Ecology* 13 (1932): 159–171.

2. Ibid., citing Herbert Howe Brancoft, *History of Utah, 1540–1886* (San Francisco: History Co., 1889), pp. 730–731.

3. USDA, Bur. Agric. Econ., Salt Lake City, Utah.

4. USDA, Bur. Agric. Econ., Div. Crop and Livestock Estimates, Utah Annual Livestock Rep., Jan. 1, 1931.

5. V. A. Young and G. F. Payne, "Utilization of 'key' browse species in relation to proper grazing practices in cut-over western white pine lands in northern Idaho," *J. Forest.* 46 (1948): 35–40.

6. Ibid.

7. Arthur W. Sampson, *Native American Forage Plants* (New York: Wiley and Sons, 1924).

8. J. S. Dixon, "A study of the life history and food habits of mule deer in California," *Bulletin Calif. Fish and Game* 30 (1934): 181–282.

9. *Range Plant Handbook* (USDA, Forest Serv., Washington, D.C., 1937).

10. Pickford, "Influence of continued heavy grazing."

11. L. A. Stoddard and D. I. Rasmussen, *Deer Management and Range Livestock Production* (Bull. 305, Utah State Agric. Exp. Stn., 1945).

12. Odell Julander and Leslie Robinette, "Deer and cattle relationships on the Oak Creek range in Utah," *J. Forest.* 48 (1950): 410–415.

13. C. M. Aldous, "A winter study of mule deer in Nevada," *J. Wildl. Manage.* 9 (1945): 145–151.

14. Ibid.

15. Nevada Department of Wildlife preliminary big game data and recommendations for hunting season dates (prepared by Mike Hess, 1993).

16. Ibid.; personal communication with Dave Mathis, retired information officer, Nevada Division of Wildlife.

17. Dave Mathis, *Following the Nevada Wildlife Trail: A History of Nevada Wildlife and Wildlife Management* (Reno: Nevada Agricultural Foundation, 1997).

18. Nev. Div. Wildl. big game status and quota recommendations 1998 (prepared by Mike Hess, big game staff biologist, Nev. Div. Wildl., Reno).

19. Personal communication with San Stiver, staff biologist with NDOW and the field biologist for the Schell Creek Range in the late 1970s.

20. Nev. Div. Wildl. big game status and recommendations 1998.

21. Personal communications, Frank Hall, area wildlife biologist, Calif. Dep. Fish and Game, September 4, 1991.

22. S. Young, "Exclosures in big game management in Utah," *J. Range Manage.* 11 (1958): 186–190.

23. J. K. McAdoo, W. S. Longland, G. J. Cluff, and D. A. Klebenow, "Use of new rangeland seedings by black-tailed jackrabbits," *J. Range Manage.* 40 (1987): 520–524; also see W. S. Longland, "Risk of predation and food consumption by black-tailed jackrabbits," *J. Range Manage.* 44 (1991): 447–450.

24. J. K. McAdoo and J. A. Young, "Jackrabbits," *Rangelands* 2 (1980): 135–138.

25. Charlie D. Clements and James A. Young, "Influence of rodent predation on antelope bitterbrush seedlings," *J. Range Manage.* 49 (1996): 31–34.

26. See, for example, R. L. Hubbard, P. Zusman, and R. Sanderson, "Bitterbrush stocking and minimum spacing with crested wheatgrass," *Calif. Fish and Game* 48 (1962): 203–208.

27. See, for example, J. P. Blaisedell and W. F. Mueggler, "Effect of 2,4-D on forbs and shrubs associated with big sagebrush," *J. Range Manage.* 9 (1956): 38–40.

28. D. N. Hyder and F. A. Sneva, "Selective control of big sagebrush associated with antelope bitterbrush," *J. Range Manage.* 15 (1962): 211–215.

29. R. B. Ferguson and J. V. Basile, "Topping stimulates bitterbrush twig growth," *J. Wildl. Manage.* 30 (1966): 839–841.

30. J. A. Young and B. A. Sparks, *Cattle in the Cold Desert* (Logan: Utah State Univ. Press, 1985).

31. C. D. Clements and J. A. Young, "A viewpoint: Rangeland health and mule deer habitat," *J. Range Manage.* 50 (1997): 129–138.

32. D. R. Dietz and J. G. Nagy, "Mule deer nutrition and plant utilization," pp. 71–78 in *Proceedings of a Symposium on Mule Deer Decline in the West* (Logan: Utah State Univ., 1976).

33. D. Griffiths, *Forage Conditions on the Northern Border of the Great Basin* (Bull. 15, USDA, Bur. Plant Industry, Washington, D.C., 1902).

34. A. S. Leopold, *A Survey of Fish and Game Problems in Nevada* (Bull. 36, Nevada Legislative Counsel Bur., Carson City, 1959).

35. B. R. McConnell and J. G. Smith, "Influence of grazing on age-yield interactions in bitterbrush," *J. Range Manage.* 30, no. 2 (1977): 91–93.

Index

Achnatherum speciosum, 9
acid detergent fiber, 139
actinomycete, 199
active restoration, 220
Adenostoma, 184
adventitious buds, 186
air layering, 63; wildfires, 185
air screen, 104
Alderfer, Jean, 55; and winter leaf retention, 152
Aldous, C. M., 16; and mule deer populations, 27
alfalfa, 183; feeding trials, 141; field deer, 155; predation, 154
alligator juniper, 49
Alnus, 201; *glutinosa*, 199
Amelanchier, 14; *pallida*, 47; *rubescens*, 15
antelope bitterbrush: achenes, 58; animals' preference for, 154; ash content, 184; browse production, 69; browsing sprouts, 188; carbohydrate reserves, 61; crude protein content, 138; deficient winter diet, 151; defoliation, 71; digestibility, leaves, 151; ecological amplitude, 194; fall and spring use, 149; figure, 18; first collected, 2; flowering, 52; flowers, 54; genetic variability, 189; grasses, 39; growth form, 69; high elevation, 43; hybridization, 12; insects, 156; Lassen, cultivar, 13; leaf chlorosis, 202; leaf hairs, 55; leaf persistence, 54; lignotubers, 189; livestock browsing, 218; mineral fertilization, 213; original establishment, 222; outcrossing, 186; perennial grasses, 218; phenology, 52; plant age, sprouting, 189; productive age, 225; range, 4; *Range Plant Handbook*, 20; required recruitment, 129; rodents, 129; second-year wood, 52; seed caches, 190; seed dormancy, 73; seedling recruitment, 135, 190; seed mass, 135; seeds, granivorous rodents' preference for, 130; seed production, 59; size classes, 63; soil moisture, sprouting, 190; spring frost, 62; sprouting, 64, 174, 189; root nodules, 198; TDN, 149; topping, 222; Warner Mountains, 15; winter range, 23
ants, 124
Apache plume, 7; figure, 19; flowers, 58; nodules, 206; phenology, 53
Arctostaphylos patula, 40; *pungens*, 49
Arcto-Tertiary flora, 10
Arno, S. F., 175
arrowleaf balsamroot, 150
Artemisia arbuscula, 43; *nova*, 47, 49
Atriplex confertifolia, 9
Axelrod, D. I., 9

Baja California, Mex., 4
Balls Canyon, Calif., 106
basalt soils, 197
Basile, Joseph, 108
basin big sagebrush, 147
Bass Hill, 37; wildfire, 185
Bayoumi, M. A., 198
Beaver Mountain deer herd, 27
Becking, J. H., 189
Bega, R. V., 198
Berg, Robert, 48; Paine Springs, 49

Bering Land Bridge, 23
bighorn sheep, 154
Big Pine, Calif., 25
big sagebrush, digestibility, 141; mule deer's preference for, 147
Bilbrough, C. J., 68
Billings, Dwight, 48, 187
Bishop, Calif., 196
Bissell, Harold, 142
bitterbrush communities: big sagebrush, 31; black oak, 36; fir, 41; greenleaf manzanita, 40; Jeffrey pine, 41; lodgepole pine, 43; serviceberry, 47; snowberry, 47; Utah juniper, 38; western juniper, 32, 35
bitterbrush tortoise scale, 168
Bitterroot National Forest, Mont., 175
black brush, 2, 50; chromosome number, 7
Blackburn, W., 47
black-tailed deer, 24
black-tailed jackrabbits, 216
Blaisdell, James, 174; reporting of bitterbrush sprouting, 66
bluebunch wheatgrass, 31, 44
Blue Mountains, Ore., 26
blue yucca, 49
Blythe Springs, Nev., 49
Boca, Calif., 189
Boise National Forest, 46
Bond, G., 199
Brassicaceae, 1
British Columbia, 2
Britton, Carlton, 186, 193
Brotherson, Jack, 50
Brown, R. C., 169
browse production, 14, 69; Dayton, 17
Buckhorn Road, 192
buckwheat, 150
Bunch, T. R., 195
bush peppergrass, 1
Butte, Mont., 188
Bylas, Ariz., 7
caches, seed, 109; nodulated seedlings, 200
calcium-phosphorus ratio, 140
California: mule deer crisis, 30; mule deer distribution, 24
captive deer, 141
carbohydrates, 138; reserves, 61
Carex rossii, 39
Carleton, J. O., 50
Carson River, 41
Carton, J. D., 24
Cascade–Sierra Nevada, 4
Casebeer, Robert, 134
caterpillars, 157
Cave Valley, Nev., 218
Ceanothus prostratus, 41; *velutinus*, 40, 201
Cedar Creek project, 220
Cedros Island, Mex., 25
cellulose, 139
Cercocarpus, 7; *Frankia* infection, 200; *ledifolius*, 83; *montanus*, 15, 41; root nodules, 200
Cervus canadensis, 26
Chaney, R. W., 9
Chapline, W. R., 20
charcoal production, 49
cheatgrass, 27, 144; ammonium sulfate, 209; Bass Hill, 38; cliffrose, 50; danger to California, 30; decoy seeds, 134; establishment of bitterbrush prior to introduction of, 224; nitrogen, 208; repeated burning, 191; root elongation, 80; seed banks, 119; seedbed quality, 123; seedling competition, 108; Washington, 45; wildfire timing, 173
Chilcoot, Calif., 201
chipmunk, 129
Chlorochroa, 162
chokecherry, 142, 144
chronosequence, soils, 197
Churchill Canyon, 153
Clark, E. C., 163

Clark, Robert, 185
cliffrose, 2, 50; achenes, 58; Arizona, 206; big sagebrush, 31; figure, 22; flowering, 63; Granite Peak, 152; hybridization, 12; hybrids, 49; leaves, 55; origin of name, 4; phenology, 52; pinyon, 50; range, 5; root nodules, 206; skunkbush, 49; Utah juniper, 32
Coleogyne, 2
Colorado Plateau, 46
Columbia Basin, 44
Columbia black-tailed deer, 25
competition index, 212
conchuela, 162
Cook, John, 188
cool-moist pretreatment, 73
conifers, 9
Cornelius, Don, 174
Cowan, James, 4
Cowania, 2; chromosome number, 7; *ericifolia*, 7; leaves, 13; *plicata*, 7; root nodules, 198; *stansburiana*, 4; *subintegra*, 7, 206
crested wheatgrass, 109
Crocker, R. L., 197
Crotalus viridis, 9
crude protein content, 138
cultivar, Lassen, 13
Curran, Mary, 4
cutworm, 168

Dahms, W. G., 42
Dallas, Ore., 59
Dalton, David, 200
damping-off, 169
Dassman, W. P., 29, 149
Daubenmire, Rexford, 31
Dayton, W. A., 16
dead fuel, 186
Dealy, J. E., 43, 122
Dearden, B. L., 150
de Candolle, Augustin, 3
deciduous shrubs, 141
decomposed granite, 195
decumbent forms, 64, 186
deer-livestock competition, 212
deer mice, 130
Desatoya Mountains, 9
desert bitterbrush, 2, 50; achenes, 58; big sagebrush, 31; black brush, 50; desert peach, 49; first collection, 3; flowers, 58; Jeffrey pine, 42; Morey Bench, 49; pinyon, 50; range, 4; root nodules, 204; tent caterpillars, 166
desert peach, 173
Desolation Ranger District, 26
Dickson, B. A., 167
diet analysis, fecal, 150; stomach, 149
digestible energy, 149
Dipodomys ordii, 129; *panamintinus*, 132
Dixie Forest, Utah, 29
Dixon, Joseph, 23, 24
Dog Skin Mountains, 111
Doman, E. R., 142
Don, David, 4
Doyle, Calif., 36, 62, 64, 78, 81, 129, 209; road kills of, 144
Dreyer, David, 82
Driscoll, Dick, 33, 185
Drivers, Charles, 175; wildfire–antelope bitterbrush relations, synthesis of, 193
Dryadeae, tribe, 7
Dryas drummondii, 199
Dyrness, C. T., 33

Eastgate, Nev., 10
Eastgate Basin, Nev., 9
Eckert, R. E., 48, 122
edaphic climax, 42
Edgerton, P. J., 43, 66
Edwards, O. T., 27
Einarsen, A. S., 27
eliasomes, 125
elk, 26, 215
elk sedge, 175
Ellison, Lincoln, 20

Elymus elymoides, 40
energy values, 140
ephedra, 174
eriophid mite, 168
Escalante, Father, 28
escape cover, 216
Eureka, Nev., 176
Evans, R. A., 91, 108, 122
Everett, R. L., 82; and decoy seeds, 134
evergreen winter browse, 151

Fallugia, 7; chromosome number, 7; leaves, 13; winter leaves, 152
Farnsworth, R. B., 198
fecal analysis, of mule deer diet, 150
fencing, to exclude deer, 215
Ferguson, R. B., 108, 195
Festuca idahoensis, 39, 43, 44
fire scars, Juniper Hill, 179
Ferguson, R. B., 108
Flukey Springs, 81, 196
Fomes annosus, 170
forage analysis, 138
Forsling, C. L., 15
Fort Ruby exclosure, 49
fourwing saltbush, 152, 183
Fox Mountain, Nev., 48
Franklin, Jerry, 33
Fraas, W. W., 188
Frankia purshiae, 200; obligate parasite, 206
Frankliniella occidentalis, 157
fuel load, of plant communities, 185
Furness, M., 156

Gamble oak, 49; digestibility, 142
Garrison, G. A., 61, 71
Gates, Dillard, 195
gelechid caterpillars, 166
gibberellic acid, 82
Glacier Bay, Ala., 199
Gondwana, 7
Granite Peak, 111
granivores, 124
grasses, native perennial: bluebunch wheatgrass, 39; Idaho fescue, 46; Indian ricegrass, 37; Sandberg bluegrass, 31; western wheatgrass, 47
grasshoppers, 168
grazing experiment, 15
Great Basin: contact time of mule deer into, 28; Experiment Station, 15; geofloras, 9; pocket mice, 129; tent caterpillar, 163
ground fires, 175

habitat type, 31
Hall, Frank, 40
halomorphic soils, 1
Hansen, R. H., 150
hard winter of 1889–90, 182
Harris, Grant, 80
harvester ants, 125
herbicidal weed control, 122
Heteromyidae, 126
Hickman, O. E., 151
Hidalgo County, N. Mex., 25
high-lined browse species, 26
Hippophae, 201
Holmgren, Ralph, 98, 119, 149, 168
Holocene, 12
Holodiscus discolor, 2
Hormay, August, 73; antelope bitterbrush range, 4; artificial seeding, 74; cache depth, 110; classic paper, 52; granivores, 124; rodent caches, 129; seed dispersal, 124; style dormancy, 74; wildfires and bitterbrush, 174
horse, evolution, 10
horsebrush, 173
Hoyle, A. E., 17
Hubbard, Richard, 69, 107, 147; mechanical site prep, 110
Hudson's Bay Company, 28
hybridization, 11

hydrocyanic acid, 3
hydrogen peroxide, 82; field preparation, 108

Idaho fescue, 27
Independence, Calif., 201
Indian remedies, 71
inoculation, 203
intergeneric fertility, 13
Intermountain area, 1
Inyo County, Calif., 62
Inyo Mountains, 25
isotopic nitrogen, 199

jackrabbit, 159; black-tailed, 216
Jackson Hole, Wyo., 45
Jacks Valley, Nev., 62
Janesville, Calif., 62
Janesville Grade, 41
jays, pinyon, 126
Jeffrey pine, 41
Jenkins, Steve, 133
Jess Valley, Calif., 15
Johnson, A. H., 184
Johnson, Tom, 186
Julander, Odell, 213
juniper: alligator, 49; California, 50; one-seeded, 49; red berry, 49; Rocky Mountain, 49; Utah, western, 35
Juniper Hill, 155, 176
juniper invasion, 177
Juniperus: deppeana, 49; *erythrocarpa*, 49; *monosperma*, 49; *osteosperma*, 49; *scopulorum*, 49

kangaroo rats, 126
Kindschy, Robert, 69
Klemmedson, J. O., 195
Krebill, R. J., 200
Kufeld, R. G., 154
Kunzea tridentata, 3; pine woodland grazing, 16

La Laguna Mountains, 25
larder-hoard caches, 126
Lassen, R. W., 144
Lassen antelope bitterbrush, 13, 170
Lassen County, Calif., 24, 36, 52, 62, 142, 176, 215; winter ranges, 23
Laurasia, 7
Lavabeds National Monument, 184
Leach, H. R., 144
Leadville Canyon, 1
leaves, winter diet of mule deer, 151
Lecanium cersifex, 168
Leman, B., 197
Leopold, Starker, 28, 29, 187, 225
LeRaye, Charles, 24
lettuce, pot test, 196
Lewis and Clark, 3, 25
lignotubers, 186
Lincoln County, Nev., 49
Linnaeus, 3
lipids, 139
Little Antelope Valley, 38
livestock reductions, 213
lodgepole pine, 200; logging, 43
Loft, Eric, 191
Longhurst, W. M., 29
Long Valley Creek, 209
low sagebrush, 43, 147; wildfires, 181
lupine, 150
Lycophotia margaritosa, 168

Madro-Tertiary geoflora, 10, 23
Magnoliales, 7
mahogany, 26, 70; feeding trials, 141
Malacosoma californicum, 162; *fragile*, 164
Malheur National Forest, 27
manzanita, 184
Martin, Robert, 186
Martinsen, C. F., 169
McArthur, E. D., 6, 8, 12, 198
McCarty, E. C., 60
McConnell, B. R., 61, 82

McNary, Ore., 59
Medicine Lake Highlands, 41, 154
Medin, D. E., 149
Mediterranean sage, 192
medusahead, 192
Meek's Table, 175
Meeuwig, R. O., 82
mesquite, 50; honey, 50
Metolius-Sisters, Ore., 59
Mica Mountains, Wash., 15
microbial fermentation, 148
minerals, 139; calcium, 140; phosphorus, 140
Miocene, 9
Modoc County, Calif., 35, 52; seeding, 110
Mohave Desert, 4
Moir, W. H., 50
Mono Lake, 42
Monsen, Steve, 119
moose, 154
Morey Bench, 11, 49
mountain big sagebrush, 47, 188; mule deer's preference for, 153
mountain brush type, 48, 188
mountain lions, 221
mountain mahogany, 7, 142; looper, 167
Mount Shasta, 35
Mueggler, Walter, 187
Muir, J. M., 200
mule deer, 23; alfalfa feeding, 141; average daily consumption, 143; browsing, 183; bucks-only hunting, 213; calorie requirements, 143; cheatgrass, 144; cheatgrass winter use, 150; doe hunts, 214; energy requirements, 198; fencing, 215; habitat created by seeding, 28; hunting season, 213; late-fall diet, 149; population counts, 214; population crashes, 213; population explosions, 27; populations at contact, 26; protection from hunting, 26; refuges, 213; stomach, 148; stomach analysis, 144; summer forage, 142; tent caterpillar leaves, 165; winter browsing, 152; winter feeding, 142
mule deer herds: Bass Hill, 37; Beaver Mountain, 27; Devil's Garden, 35, 153, 196; Entiat, 150; Kaibab, 27; Lassen, 225; Lassen-Washoe, 144, 191; Murders Creek, 27; Piceance Basin, 150; Silver Lake, 43
mule deer subspecies: Burro, 25; California, 24; Cedros Island, 25; Peninsula, 25; Rocky Mountain, 24; Tiburon, 25
Munns, D. N., 201
Murray, Robert, 191
mutualistic association, 129
mycorrhizal, 208

native perennial grasses, 117
needlegrass: desert, 9; needle-and-threadgrass, 45, 49; Thurber's, 35; western, 38
Nelson, David, 169, 206
Netz, Charles, 72
nitrapyrin, 209
nitrification, 208
nitrogen: carbon immobilization, 210; enrichment, 196
nitrogen fixation, 205
nocturnal seed caches, 126
Nord, Eamor, 50, 169; bitterbrush ash content, 186; bitterbrush sprouting, 64, 124; classic paper, 52; desert bitterbrush, 42; excised embryo test, 73; granivores, 124; rodent caches, 129; seed collection, 99; seed dispersal, 124; viability test, 102
nutcrackers, 126

oak: California black, 37; Gamble, 49; Oregon, 35; waveyleaf, 49
Observation Peak, 192
oceanspray, 2
Ochco Mountains, Ore., 40
Odocoileus hemionus subsp.: *californius*, 24; *cerrosensis*, 24; *columbianus*, 24;

crooki, 24; *eremicus*, 24; *fuliginatus*, 24; *inyoensis*, 24; *peninsulae*, 24; *sitkensis*, 24
Ord's Kangaroo rat, 129
Orgyia vetusta gulosa, 168
Owens Valley, 201
Owyhee Mountains, 35, 46

Pacific County, Wash., 25
Paine Springs, Nev., 49
Paisley Ranger District, 116
Paleogene, 7
Panamint kangaroo rat, 132
passive management, 227
Paulina Mountains, Ore., 59
Pechanec, Joe, 174
Peninsula mule deer, 25
perennial grass fuel, 172
Person, Bennett, 80
Peterson, R. A., 80, 169
phenolic compounds, 82
Phigalia plumogeraria, 166
Phillips, Bill, 165
Philpot, C. W., 186
phosphorus deficiency, 154
Pickford, G. D., 211
pine: lodgepole, 42; ponderosa, 39
pine/bitterbrush woodlands, 2
pinegrass, 175
Pinus: cembroides, 48; *contorta*, 42–43; *monophylla*, 48
pinyon, 48; encroachment, 215; spread, 12
Pit River, 41
Pittack, C. C., 197
Pleistocene, glaciation, 23
Plumas County, Calif., 38
pluvial lakes, 1
pocket mice, 126
Pogonomyrmex, 124
ponderosa pine, 2; fire scars, 175; wildfires, 174
post cutting, 181
pot-testing soils, 196
prescribed burning, 186
preserves, wildlife, 221
Price, R., 60
Pringle Falls, Ore., 67
promiscuous burning, 176, 181
pronghorn, 154
proximate analysis, 139
Prunus andersonii, 49
Pseudotsuga menziesii, 41
Psorothamnus, 1
pumice soils, 200
Purpus, J. A., 11
Pursh, Frederick Traugott, 3
Purshia: bitter taste, 3; chromosome numbers, 7; evolution, 12; hybridization, 11; leaves, 13; livestock forage, 211; *mexicana* var. *stansburiana*, 2; natural hybrids, 12; seedling predation, 215; *tridentata*, 2; *tridentata* var. *glandulosa*, 2
pygmy conifer, 1

Quercus: gambelii, 49; *garryana*, 35; *grisea*, 49; *kellogii*, 36

rabbitbrush, 173
Rafinesque, C. S., 24
Range Plant Handbook, 20
Rasmusson, D. I., 142, 212
red berry juniper, 49
refugia, 2
Reynolds Creek, 46
Rhus trilobata, 49
Ribes cereum, 43
Rice, Carol, 184
Richards, J. H., 68
Rickard, W. H., 45
Riffle, J. W., 82
Righetti, Timothy, 201
road kills, 144
Robbinette, Leslie, 213
Robertson, J. H., 107
robins, 182
Rock Creek, 47

Rocky Mountain(s): antelope bitterbrush range, 4; juniper, 44; mule deer distribution, 24; National Park, 41
rodents: mutualistic association, 129; seed nutrition, 131; seed predation, 217; tetramine-treated seeds, 134
root-nodulated genera, 198
Rosa, 15
Rosaceae, 7; genera, 7
Rosideae, subfamily, 7
Rummell, Robert, 175
Russian thistle, 106

sage, 9; Lahontan, 38
sagebrush: animals' preference for, 221; big, 31; black, 49; competition, 221; excessive utilization, 221; family, 2; low, 38; mountain, 48; rock, 46; silver, 46; succession, 174; three-tip, 186
saltbushes, 1
Sampson, Arthur W., 14, 211
Sanderson, Reed, 69
Sauer, R. H., 45
Say's stinkbug, 162
scalper, 120
scatter-hoard caches, 126, 130
Schell Creek Range, 214
Schnegas, Edward, 126
Schussler, Howard, 126
Schuttler, Patti, 206
Sciuridae, 126
seed: dormancy, 73; snow cover, 75; stratification, 73; style, 74; tetramine, 134
seedbed temperatures, 88, 91, 96–97
seedbed pathogens, 170
seed collection: air screen, 104; damaged seeds, 102; hoop and bag, 100; moisture content, 101; threshing, 102; vacuum harvesters, 100
seed germination: embryo oxygen, 78; fall seeding, 76; hydrogen peroxide, 82; optimum temperature, 88; osmotic potentials, 78; stratification duration, 80; thiourea, 80
seeding, 98; aerial, 111; caches, 109; depth, 111; drill calibration, 112; evaluation success, 113; microsites, 107
seed orchards, 166
seed production, 98
seedling competition, 113
serpentine soils, 35
serviceberry, 14, 70
Seven Lakes Range, 38
shadscale, 9, 199
Shantz, H. L., 16
Shaw, Nancy, 52, 169
Shepherdia, 201
shrub: architecture, 68; steppe, 1
Sierra-Cascade, 1
Sierra Madre Oriental, 7
Sierra Valley, Calif., 201
Silver Lake, Ore., 122, 151
single-leaf pinyon, 48
Siskiyou County, Calif., 35
Sitka deer, 25
skunk-bush, 49
Smith, A. D., 147, 198; on sagebrush digestibility, 141
Smith, Dixie, 45
Smith, Jedediah, 28
Smith, Ralph, 214
Smothers, G. A., 184
Snake River Plains, 59, 174
Sneva, F. A., 193, 221
snowbush, 145, 201
soil-plant relations, 195
soil potential, 225
soil-surface germination, 75
Southern mule deer, 25
Sparhauk, W. N., 16
Spencer, D. A., 133
Sprengel, Kurt, 3
Springfield, H. W., 82
spring frost, 62
sprouting bitterbrush, 174; origin, 194
squaw carpet, 144, 163
stand renewal process, 171
Stansbury, Capt. Howard, 4

Stansbury Island, 4
Stanton, Frank, 59
Stebbins, G. L., 11
Steep Mountain, Mont., 188
stocking rate, bitterbrush, 118
Stoddart, L. A., 212
Storm, E. V., 15
stratification, 73
strong fat, 20
strong feed, 15
Stutz, Howard, 12
sulfur deficiency, 196
symbiotic fixation, 197
Symphoricarpus: albus, 41; *longiflorus*, 47

temperature profiles, 89–91
tent caterpillars, 162; detriment to browsing, 165; epizootic virus, 166; wildfires, 166
Tertiary vegetation, 10
Tew, Ronald, 46
thiourea, 80; spring planting, 81; toxicity, 81
Thomas, L. K., 12
Thompson, Clarence, 166
three-tip sagebrush, 157
Tiburon Island, 25
Tiedemann, A., 198
Tigarea tridentata, 3
topoedaphic climax, 42
Torrey, John, 4
Train, Perry, 71
Trinity County, Calif., 4, 35
Trousdale, Eugene, 82
Truckee, Calif., 2
Tueller, P. T., 31, 46, 48, 149
Tule Peak, Nev., 38

Urness, P., 154
Utah: early mule deer population, 29; first hunting season, 29; Great Salt Lake, 4; juniper, feeding trials, 141; serviceberry, 151

Valmis, J., 197
Vander Wall, Steve, 126
Vauquelinia, 10
vegetation: classification, 16; manipulation, 225
viability test, 102
Victoria Mountains, Baja Calif., Mex., 25
vitamins, 132, 140
Volland, L. A., 42

Wagle, R. F., 80, 196
Wagstaff, Fred, 188
Wallowa Mountains, 43
walnut spanworm, 166
Warm Springs, Ore., 59
Warner Mountains, 15
Wasatch County, Utah, 188
waveyleaf oak, 49
Weaver, Harold, 175
Webster, S. R., 199
Welch, Bruce, 148, 152, 155
West, N. E., 39, 46
western juniper, 26, 195; Cedar Creek, 220; fruits, 181; seedling establishment, 182; segregating preference for by mule deer, 152
western tussock moth, 168
wheatgrass: bluebunch, 45; crested, 181; intermediate, 181; western, 47
White Mountains, 25
White Pine County, Nev., 27
White Pine deer herd, 27
white-tail deer, 23; destroyed population, 26
white-throated wood rats, 133
wildfires: catastrophic, 219; clipping interaction, 193; energy, 186; frequency, 175; grazing management, 191; ground, 172, 175; Native Americans, 172; natural environment, 171; promiscuous burning, 224; shrub understory, 175; species ecology, 176; sprouting species, 173; suppression, 171; timing, 172

Wildland Seed Laboratory, 74
wild rose, 1, 149
Williams, Stephen, 208
Willow Creek, Calif., 69
winterfat, 139
winter leaves, protein, 152
Wolland, A. G., 199
Work, John, 28
Wright, Henry, 176
yellow star thistle, 192
Youngberg, C. T., 42, 199
Yucca baccata, 49

Zobel, Donald, 200